Postmodern Wetlands

POSTMODERN THEORY

Series editor:
THOMAS DOCHERTY
School of English, University of Kent at Canterbury

This series openly and vigorously confronts the central questions and issues in postmodern culture, and proposes a series of refigurations of the modern in all its forms: aesthetic and political, cultural and social, material and popular. Books in the series are all major contributions to the re-writing of the intellectual and material histories of socio-cultural life from the sixteenth century to the present day. They articulate or facilitate the exploration of those new 'post-theoretical' positions, developed both inside and outside the anglophone world – inside and outside 'western' theory – which are pertinent to a contemporary world order.

Other titles in the series include:

After Theory
Thomas Docherty

Justice Miscarried: Ethics, Aesthetics and the Law
Costas Douzinas and Ronnie Warrington

Towards a Postmodern Theory of Narrative
Andrew Gibson

Jarring Witnesses: Modern Fiction and the Representation of History
Robert Holton

POSTMODERN THEORY

Postmodern Wetlands
Culture, History, Ecology

RODNEY JAMES GIBLETT

Edinburgh University Press

© Rodney James Giblett, 1996

Transferred to digital print 2007

Edinburgh University Press
22 George Square, Edinburgh

Typeset in $9\frac{1}{2}$/12 point Melior by
Photoprint, Torquay, Devon and
printed and bound in Great Britain
by CPI Antony Rowe, Eastbourne

A CIP record for this book is available
from the British Library

ISBN 0 7486 0844 3

The right of Rodney James Giblett to be identified as
author of this work has been asserted in accordance
with the Copyright, Designs and Patent Act (1988)

'Greasy Lake', from T. Coraghessan Boyle, *Greasy Lake
and Other Stories*, copyright © 1979, 1981, 1982, 1983,
1984, 1985 by T. Coraghessan Boyle. Used by permission
of Viking Penguin, a division of Penguin Books USA Inc.

When I would recreate myself, I seek the darkest wood, the thickest and most interminable and, to the citizen, most dismal swamp. I enter a swamp as a sacred place, a *sanctum sanctorum*. There is the strength, the marrow, of Nature.

Henry David Thoreau

Contents

List of Illustrations

Preface

Wetlands – swamps, marshes, mires, morasses, bogs, lagoons, sloughs, shallow lakes and estuaries, etc. – have been seen by many in 'the west' as places of darkness, disease and death, horror and the uncanny, melancholy and the monstrous – in short, as black waters. Yet a minority in 'the west' and the indigenes of Australia (and no doubt elsewhere) regard wetlands as places of both life and death, light and dark, as biologically rich and fertile, mucky and murky, vital for life on earth – in other words as living black waters.

These viewpoints connect historically and engage politically with two further traditions, or moments, and practices: the latter with matrifocal gylany[1] which feminised the swamp positively as the source of new life in the Snake Goddess, the mistress of living black waters; and the former with patriarchal hierarchy with its dryland agriculture and its misogynist denigration of the wetlend as the environmental *femme fatale*, spider woman and *vagina dentata*. With the rise of capitalism under the aegis of patriarchy in Europe with its modern cities the black waters of wetlands 'at home' and in the colonies were seen by many citizens as pre-modern wasteland or wilderness to be conquered as a marker of 'Progress'. Wetlands either were drained or filled to create the dead surface of private property on which agricultural and urban development could then take place or they were polluted by cities and farms to produce the dead black waters of a modern waste-wetland. Later the draining or filling and polluting of wetlands by industrial technology increased markedly their degradation and destruction.

Yet with the rise of the conservation movement, ecology and green politics the conservation or rehabilitation of postmodern wetlands as fully-functioning ecosystems and habitats harks back to the living black waters of pre-modern, pre-capitalist, matrifocal wetlands. This book plots the contention between these viewpoints and positions, and takes sides in the struggle between them. The position enunciated in this book is highly critical of the view of wetlands as black waters and of the draining, filling or polluting of them. It is unashamedly in solidarity with the view and tradition of conserving wetlands as living black waters. The ethics and practices of wetlandcare associated with this view and tradition have only recently started to conserve or rehabilitate wetlands as vital for life on earth. This book takes sides with the wetlands of the world, or what's left of them – living or dead, living and dying.

Forrestdale,
July 1994 and December 1995

NOTE

1. Defined by Riane Eisler (1987, pp. 165–6) as 'the social structure in which both sexes were equal'.

Acknowledgements

Many people have helped in the writing of this book. Often this has been by way of passing helpful references on to me about the representation of wetlands in various places, or by suggesting profitable lines of research. At other times it has been by way of offering helpful comments on my fledgling, but failing, theoretical insights, especially in the early stages of this project. I would like to thank specifically Maria Angel, Betty Benjamin, Ron Blaber, Anne Brewster, Bethany Brown, Marion Campbell, Andrew Del Marco, Brian Dibble, Rita Felski, John Fielder, Helen Flavell, Sandra Giblett, Peter Gilet, Cheryl Gole, Yvette Grant, Susan Hayes, Alan Hill, Bob Hodge, Christina Lupton, Ann McGuire, Alec McHoul, Margaret MacIntyre, Ann-Marie Medcalf, Toby Miller, Stephan Millett, Karl Neuenfeldt, Constance Ngin, Wendy Parkins, Ned Rossiter, Graham Seal, Zoë Sofoulis, Jon Stratton, Jim Warren and Hugh Webb for their help. Any shortcomings in this book, any failures to follow up or to respond to their suggestions and comments, are entirely my own fault.

I would also like to make a special acknowledgement of the assistance of the staff of the Inter-Library Loan section of the Robertson Library of Curtin University, especially Barbro Hicks, who assiduously pursued often obscure references and invariably obtained hard-to-get materials. In addition I would like to make a general acknowledgement of the support of Sandra Giblett who has participated in this project since its inception and who has tried to help me to keep my feet on the ground, or in the wetland, and my head out of the clouds of the heady heights of 'theory'. I am also grateful to the Indian Ocean Centre for Peace Studies for research assistance for Chapter 3. Last, but by no means least, Thomas Docherty and Jackie Jones have always provided helpful and encouraging editorial advice.

Grateful acknowledgement is made to the following publishers for permission to reproduce copyright material: Victor Gollancz Ltd for *The Drowned World* by J. G. Ballard; Viking Penguin Inc. for *Greasy Lake and Other Stories* by T. Coraghessan Boyle; and Macmillan Ltd, publishers of Picador Books, and A. P. Watt Ltd for *Waterland* by Graham Swift.

Aesthetics and the Wetlandscape

Introduction: Where Land and Water Meet

Wetlands are not always, and for some not ever, the most pleasant of places. In fact, they have often been seen as horrific places. In the patriarchal western cultural tradition wetlands have been associated with death and disease, the monstrous and the melancholic, if not the downright mad. Wetlands are 'black waters'. They have even been seen as a threat to health and sanity, to the clean and proper body, and mind. The typical response to the horrors and threats posed by wetlands has been simple and decisive: dredge, drain or fill and so 'reclaim' them. Yet the idea of reclaiming wetlands begs the questions of reclaimed from what? for what? for whom? A critical history of wetlands' drainage could quite easily be entitled 'Discipline and Drain'.

But why this horror of wetlands? Part of the problem lies in the fact that wetlands are neither strictly land nor water. Rather, they are both land and water. Wetlands often represent 'a temporal and spatial transition from open water to dry land' (Niering 1991, p. 21), as well as representing a temporal and spatial mediation between the two, what could be called 'the quaking zone'. The waters of wetlands can even be in temporal transition in their spatial locality: '*swamp* is when the water goes in one end and out the other, *bog* is when it goes in and stays in' (Atwood 1991, p. 87). If not in transition, many wetlands are 'physically halfway between the water and the land . . . Bogs are a different kind of halfway world, neither water nor land yet a part of both' (Coles 1989, p. 151).

Physically wetlands are bad enough but morally wetlands are the modern environmental *demi-monde* with all the feminised

3

horrors of the half-known life which accrue in conventional minds to the modern urban *demi-monde*. Walt Whitman summed it up in 1860 when he referred to 'the strange fascination of these half-known/half-impassable swamps, infested by reptiles/ resounding with the bellow of the alligator, the sad noise of the rattlesnake' (Miller 1989, p. 60). Wetlands are a place of the alien, reptilian 'other', even the home of monsters lurking in their murky depths. Rather than fascination, horror has been the typical patriarchal response to wetlands which have been seen as infested with malaria, miasma and melancholia.

More generally, wetlands are an anomaly in a classificatory order predicated on a hard and fast distinction between land and water, time and space, or perhaps more precisely, their representational systems: the timelessness of maps and the spacelessness of history do not lend themselves to the changing nature of wetlands. What is needed instead are temporal maps that move with time, that show historical change in wetlands; spatial history (Carter 1987, intro.) that shows history taking place in, and in relation to, wetlands; and quantum ecology that construes the natural environment, especially wetlands, on a space/time continuum.

THE BLACK WATERS OF WETLANDS

The wetland figured as 'black water' with all its incipient racist associations would be a convenient short hand label to encapsulate more than two millennia of patriarchal western vilification and destruction of wetlands. An exemplary instance of this figuring is found in Wilkie Collins' *The Woman in White* of 1859–60 set mainly in Blackwater Park with its Blackwater Lake. Marian Halcombe takes a walk in the park and finds herself:

standing suddenly on the margin of a vast open space, and looking down at . . . Blackwater Lake . . . The lake itself had evidently once flowed to the spot on which I stood, and had been gradually wasted and dried up to less than a third of its former size. I saw its still, stagnant waters, a quarter of a mile away from me in the hollow, separated into pools and ponds by twining reeds and rushes, and little knolls of earth. On the farther bank from me the trees rose thickly again, and shut out the view, and cast their black shadows on the sluggish, shallow water. As I walked down to the lake, I saw that the ground on its farther side was damp and marshy, overgrown with rank grass and dismal willows. The water, which was clear enough on the open sandy side, where the sun shone, looked black and poisonous opposite to me, where it lay deeper under the shade of the spongy banks, and the rank overhanging thickets and tangled trees. The frogs were croaking, and the rats were slipping in and out of the shadowy water, like live shadows themselves, as I got nearer to the marshy side of the lake. I saw here, lying half in and half out of the water, the rotten wreck of an old overturned boat, with a sickly spot of sunlight glimmering through a gap in the trees on its dry surface, and a snake basking in the midst of the spot, fantastically coiled and treacherously still. Far and near the view suggested the same dreary impressions of solitude and decay, and the glorious brightness of the summer sky overhead seemed only to deepen and harden the gloom and barrenness of the wilderness on which it shone. I turned and retraced my steps to the high heathy ground ... (Collins (1859–60) 1974, pp. 227–8)[1]

This passage reproduces much of the traditional western iconography and phenomenology of the wetland. The marshy side of the lake is figured as a kind of Eden-after-the-Fall in which a satanic serpent lurks. Later we shall see that Dante and Milton had gone even further than Collins in satanising the swamp as part of their theologising of the landscape: Dante by figuring one circle of hell as a slimy Stygian marsh, and Milton by troping Satan as a monstrous swamp serpent who is generated out of the slime of hell.

What is probably more remarkable, though, about this passage from *The Woman in White* than its reproduction of traditional Christian iconography, and its moralisation of the marsh, its *marécage moralisé*, is the trajectory it traces through space from the high heathy, even healthy, ground descending spatially and morally to the low poisonous, decidedly unhealthy wetland, and the juxtaposition it sets up between the sublime sky and the slimy swamp; between the openness of the former and the claustrophobia of the latter; the life and light of the former and the death, darkness, decay and disease of the latter; between the glories of the former and the wasteland of the latter.

The narrator/character of Marian Halcombe is positioned as first person in relation to the landscape, or more precisely what could be called the wetlandscape, or at least its visual space, entirely via the sense of sight. The repetitious insistence on, and of, 'I saw' places her in relation to various 'views', or lack of them, the landscape reduced to occasions for taking, or thwarting, 'views'. She is not represented as hearing, touching, smelling or tasting the wetlandscape but is placed in a position of mastery from which to look down on the wetland in order to know it and so master it, even though it is waste land. In fact, the 'I' is arguably consti- tuted as bourgeois, indeed petit bourgeois, individual, albeit masculinised, subject insofar as the waste/wet land is postulated as alien other.

The figuring of the wetland as wasteland is exacerbated for Marian one windy and cloudy morning when 'the rapid alterna- tions of the shadow and sunlight over the waste of the lake made the view look doubly wild, weird, and gloomy' (ibid. p. 253). The surface of the water reflects and 'heightens' the depths of gloom of the sky on a dark and overcast day, whereas on a bright and glorious day it is juxtaposed with it. The surface of water and the depths of the sky are set up with the petit-bourgeois individual consciousness situated between, and mediating, them. Philo- sophers may endlessly debate whether a tree falling in a forest occurs if there is no individual consciousness to perceive it, but for there to be a reflection in water there has to be a visually perceptive individual consciousness to receive it.

Indeed, the individual consciousness is constituted in modern European landscape painting not only as the point of intersection of the lines of perspective, but also as the terminus of the lines of reflection. The surface of water, Henri Lefebvre has argued, 'symbolises the surface of consciousness and the material (concrete) processes of decipherment which brings what is obscure forth into the light' (Lefebvre 1991, p. 86). Consciousness deciphers the marks on the surface of the water and brings the obscure into transparency, dark into light. Yet for there to be this process of enlightenment, there has to be a corresponding process of 'endarkenment'. The individual process, state and cultural mo(ve)ment of enlightenment is linked to and made possible by concomitant acts of 'endarkenment', whether it be of the so-called dark continent of Africa or of female sexuality (or perhaps more precisely the relation to the mother[2]), or of the 'black water' of the wetland, all of which tend to get conflated and used to figure each other in the patriarchal western tradition.

The reduction of the water to surface for decipherment also denies, if not represses, the depths of the water. Water, Luce Irigaray argues, 'serves as a reflecting screen and not as a reminder of the depths of the mother; it sends back the image of the sun, of men, of things . . .' (Irigaray 1985, p. 289). The individual consciousness reflecting on the surface of things is a patriarchal consciousness which represses the depths of the mother. It necessarily finds the wetland boringly repetitive unless it is broken up by pools of reflective water. In Rider Haggard's *She* there were 'wide stretches of lonely, death-breeding swamp, unbroken and unrelieved so far as the eye could reach, except here and there by ponds of black and peaty water that mirror-like, flashed up the red rays of the setting sun' (Rider Haggard (1887) 1991, pp. 65–6).[3] The patriarchal consciousness is here able to situate itself in relation to the datum point of the sun via the occasional reflective pool of albeit black water, in which what Julia Kristeva calls 'the black sun of melancholia' (Kristeva 1989) is reflected. Without these markers it/he would be in danger of being lost in the depths of the swamp monster/mother.

Yet when the individual consciousness is set up in this reflexive, or more precisely self-reflexive relationship between water and sky, it is constituted as narcissistic, and as self-seduced, paranoiac of absorption by the mother's body, or by bodies of water. The mirror of water, Jean Baudrillard argues, is 'not a surface of reflection, but of absorption . . . It is always a matter of self-seduction . . . All seduction in this sense is narcissistic' (Baudrillard 1990, p. 69).

THE PICTURESQUE SWAMPS?

Given the horrors of the depths of the swamp to the patriarchal consciousness it is only 'natural' or inevitable that the wetland wasteland is quite useless from a capitalist agricultural point of view, though it may be picturesque from an aesthetic point of view and yet at the same time the perfect setting for a murder, and so associated with an undesirable and illegal death, three facts which are associated in Sir Percival's mind in *The Woman in White* when he relates how:

'Some people call that picturesque,' said Sir Percival, pointing over the wide prospect [of the lake], with his half-finished walking-stick. 'I call it a blot on a gentleman's property. In my great-grandfather's time the lake flowed to this place. Look at it now. It is not four feet deep anywhere, and it is all puddles and pools. I wish I could afford to drain it, and plant it all over. My bailiff (a superstitious idiot) says he is quite sure the lake has a curse on it, like the Dead Sea. What do you think, Fosco? It looks just the place for a murder, doesn't it?' (Collins (1859–60) 1974, p. 253)

For Sir Percival the wetland is a blot, a black mark, on a gentleman's property, on its clean and proper surface, on his character, on its title deeds and in its ledgers, and not even a debit in red. Sir Percival refers to this blot in the twin indexical gestures of the demonstrative 'that' (with no following substantive as if it were indescribable) and pointing with his stick as if he were unable to describe it adequately, or even get it into words.

Indeed, the wetland constitutes the unutterable in the patriarchal western tradition, hence the recourse that is invariably made, and Marian Halcombe as a narrator of *The Woman in White* is no exception, to the lexicon of what could be called standard swampspeak with its vocabulary of such well-worn words as 'dismal' (ibid. p. 279), 'dreary' (ibid. pp. 228, 253 and 280), 'desolate' (ibid. p. 280), 'gloomy' (ibid. p. 253) and so on. Wetlands, perhaps more than any other site, bear out Jacques Derrida's contention that 'we find ourselves unable to represent 'place' itself except by metaphors' (Derrida 1981, p. 160). Wetlands have almost invariably been represented in the patriarchal western tradition in metaphors of despair and despondency in an overworking of the nether regions of the psychopathological register and of the lower echelons of the pathetic fallacy in which the psychological is projected on to the geographical.

Equally we find ourselves unable to represent states or abstractions, or at least our stance towards them, except by metaphors of place or sites. Often our stance towards an ideological or political position is figured in terms of our stance towards a place. Wetlands have been used metaphorically to convey our repudiation or vilification of the ideologically incorrect in an overemphasis on the bass notes of the psychopolitical scale. A favourite expression in the vocabulary of recent critical theory is the word 'mire', either as noun or verb.[4] Capitalism, colonialism or patriarchy, for example, are figured as a mire. To take an ideological *faux pas* and to be mired in one of them, or what's worse in a combination of two of them, or even all three (horror of horrors), is to find oneself sunk irretrievably off the straight and narrow of the ideologically correct. Yet the pejorative use of the mire metaphor is implicated ('mired') in the capitalist, colonialist and patriarchal repudiation and vilification of wetlands. Such theorists would never dream of using a racist or sexist metaphor to figure the ideologically incorrect, but persist in the use of 'placist' metaphors of wetlands and so are what could be called 'mis-aquaterrists,' haters of wetlands.

Besides recourse to the merely indexical, Sir Percival has recourse to the imperatives of capitalist agriculture which dictate that the wetland is useless land as it stands and that it should be drained and transformed into productive land. For Sir Percival water under four feet deep, a rough rule of thumb for defining the upper limit of water in a wetland, is by definition useless, as are puddles and pools, almost too shallow for a wetland except perhaps for seasonal or temporary wetlands which could even be dry for much of the year. The wetland could be viewed as picturesque, but never as beautiful, and certainly not ever as producing the sublime, the pinnacle (the mountain metaphor is appropriate) of aesthetic pleasure, albeit bordering on pain in the case of the sublime, the upper echelons of the psychogeographical register and the higher notes on the psychopathological scale. At the other end of the scale, wetlands constitute, by and large, the obverse, or secret as Zoë Sofoulis (1988) puts it, of the sublime which she construes as the slimy.[5]

Aesthetics and wetlands have had a fraught and troubled relationship. Two wetlands' scientists have tried to 'talk up' the aesthetic status of wetlands by claiming that 'the aesthetics of a landscape where water and land [are found] together often provide a striking panorama' (Mitsch and Gosselink 1986, p. 4). The implication of the qualification 'often' is that wetlands sometimes do not, and traditionally they have not at all. The only picturesque wetland is artificialised wet land, the canalised, tamed and straitened river such as in the paintings of Constable, or the ponds of 'entry statements' to housing estate developments. It is no accident, as Anne Bermingham points out, that the enclosure of the English countryside, much of which took place in wetlands, occurred at the same time as the emergence of rustic landscape painting as a major genre in England, much of which in Constable's case was of enclosed wet lands (Bermingham 1986, p. 1).[6] The narrative closure of rustic landscape painting went hand in hand with the agricultural enclosure of common (wet) land as private property. These two processes of what Bermingham calls 'the aesthetics of the painted landscape and the economics of the enclosed one' were linked by the concept of 'Nature'. The picturesqueness of these landscapes enabled the individual patriarchal consciousness to reflect upon itself in reflecting on 'Nature' via the surface of still water.

Constable's landscapes, or more precisely his artificialised wetlandscapes, Bermingham has argued, 'represent an important encoding of one myth of bourgeois realism which held that meaning and truth could be read off the surface of things . . .' (Bermingham 1990, p. 112). The fact that the paintings often depicted reflections in the water (usually of the sky) is no accident as the painting thus sets up a position for the viewer between the still surface of the water and the sublime sky as the point of reception for the reflection. From this position the viewer could reflect on the surface of the beautiful components of the English pastoral landscape (horses, trees, canals, locks, men, etc.) as objects of ownership, as private property. Indeed, English pastoral landscape as depicted by Constable was both the repository of national identity and of private property composed of objects of ownership with the former predicated on the latter. What is more, the paintings were situated between, and framed out, the slimy depths of the wet land hell below the surface of the water whilst depicting as mere background the sublime heights of the heavens usually depicted as billowing, even mountainous, clouds. In the process, water was denied depth and reduced to surface.

The wetland could only be picturesque at a distance amongst other things because the picturesque is, argues Paul Carter, 'that which pleased the curious eye . . . Picturesque prospects were ones that allowed the eye to wander from object to object.' Above all, the picturesque delineated 'a place for travelling,' it was 'a traveller's viewpoint, a possible stopping place' (Carter 1987, pp. 231–2 and 254).[7] But a place necessarily to hold at arm's length, to master, and even own, as object, as property, to frame within the picture, and to situate oneself in between the water and the sky.

Generally both 'at home' and 'abroad' in the colonies wetlands were only picturesque when viewed from a distance, and preferably from a raised vantage point. Explorers of the Swan Coastal Plain of Western Australia in the early 1830s related how: 'nine miles from the lake we left in the morning, an agreeable and sudden change took place in the scenery; we had almost imperceptibly ascended an eminence commanding an extensive view of a vast plain, bounded to the eastward by a range of majestic mountains . . .' (*Journals* (1833) 1980, p. 2).[8] In other words, the lake itself, probably Forrestdale Lake, was not agreeable scenery, but a vast plain was when viewed from a position of pre-eminence, if not mastery, bounded by 'a range of majestic mountains'. 'Vast plains', even if interspersed with lakes, could be picturesque, but not views of the lakes themselves.

Wetlands are generally not picturesque as Sir Percival indicated. Quite the contrary. In Australian travel literature, Carter goes on to argue, 'picturesqueness was a quality ascribed to two very different kinds of countryside. On the one hand, it was applied to 'grassy meadow' and extensive 'plains'; on the other, it was thought appropriate to regions of 'lofty mountains', 'impervious thickets' and winding streams' (Carter 1987, p. 232). And certainly not to 'dismal swamps' or 'black waters', scenes which did not conform to the ideal of the pastoral landscape of the English countryside.

The confines of closed, wooded wetlands, such as swamps, pose a problem for an aesthetics of sight which is accustomed to seeing without interruption or impediment whereas the seeming limitlessness of open, grassed wetlands, such as marshes, are inimical to taking the long view within confined borders, down tree-lined avenues, bounded by mountain ranges. Rider Haggard's narrator in *She* repeatedly harps on 'measureless swamps' (Haggard (1887) 1991, p. 28), on 'endless desolate swamps that stretched as far as the eye could reach' (ibid. p. 61), on the fact that 'the swamp was apparently boundless' (ibid. p. 63), on 'that measureless desolation' (ibid. p. 66), on 'a region of almost endless swamps' (ibid.

p. 73), on 'the region of eternal swamp' (ibid. p. 76), on 'a vast extent of swamp' (ibid. p. 91) (not even accorded the definite article), on 'miles on miles of quagmire . . . miles on miles of it without a break' (ibid. p. 116) and on 'the boundless and melancholy marsh' (ibid. p. 314). Wetlands 'do not cater to established classical concepts of vista, horizon and landscape' (Fritzell 1978, p. 530). In fact, they are quite resistive to delineated vista and limited horizon. The wetlandscape is inimical to landscape. Only when they are placed in the midst of 'a vast plain' bounded by mountains can wetlands become picturesque.

Yet some writers have tried to portray the wetland in less pejorative terms than Haggard, and have even attempted to develop a counter-aesthetics of the wetlandscape. The narrator of Sarah Orne Jewett's *A Marsh Island* first published in 1885 appreciated the limitless vista of 'the marshes [that] seemed to stretch away to the end of the world' (Jewett 1885, p. 3). Whereas in *She* the Haggardian narrator feels that the swamps defy the sense, and power, of sight, because they go beyond its reach, are outside its purview, in *A Marsh Island* the narrator sees the marsh as a limitless horizon of possibilities. Or if not limitless, like the wetlands of Warrick Wynne's poem with their 'ground like soaked sponge, the limitless sky', they could be seen as at least having an 'extended horizon' (Wynne 1992, pp. 17 and 18). The wide horizons of fens and marshes, the repetition *ad infinitum*, even *ad nauseam*, of swamps, imply monotony to the patriarchal consciousness, though they can connote a wide horizon of possibilities.

The cultural rehabilitation of wetlands being attempted in this book would involve seeing them as valuable, or even pleasing, without their necessarily being seen as beautiful, picturesque or sublime in conventional terms. Yet wetlands may not be regarded as ecologically valuable until they are seen as aesthetically pleasing. To see them as pleasing would entail rethinking the whole category and function of the aesthetic, especially of what constitutes a landscape. In a paper entitled appropriately 'Scapeland', Jean-François Lyotard argues that 'deserts, mountains and plains, ruins, oceans and skies enjoy a privileged status in landscape painting' (Lyotard 1991, p. 184). Wetlands, on the other hand, do not enjoy such a status in landscape painting. Wetlands are more 'scapeland' than landscape, the geographical equivalent of the scapegoat on which communal sins are heaped. The scapewetland is then driven out to die in the wilderness and so the sins of the community are expiated, the community cleansed of its moral dirt by constituting the wetland as dirt, as 'matter out of place', to use Mary Douglas' terms (Douglas 1966, p. 35). Wetlands, as we shall see in later chapters, are water in the place of land and darkness in the place of light.

Swamps did enjoy a privileged status and a brief moment of flowering in American landscape painting and literature during the so-called American Renaissance of the nineteenth century, but this seems to be the exception which proves the rule (Miller 1989). By and large swamps have never even had a place, let alone a privileged place, in landscape painting. But as Harry Godwin said, 'any fool can appreciate mountain scenery. It takes a man [sic] of discernment to appreciate the Fens' (Coles 1989, p. 58). Wetlandscape photographs, such as those by Bill Thomas, or pen and ink drawings of wetland scenes, such as those by Albert Hochbaum, are not necessarily, and are not usually considered as, beautiful, picturesque or sublime.[10] Yet these three are not the only aesthetic modes possible. Freud pointed to a fourth possibility, the uncanny, the *unheimlich* or unhomely which applies precisely to the wetlandscape.

The uncanny combines fascination and horror. These sensations are produced more by the immediate and visceral senses of smell, taste and touch than by the distancing and masterful senses of sight and hearing. As the wetland is a place which assualts the senses of smell and touch, more so than the other senses, it is the uncanny place par excellence which gives rise to fascination and horror. Yet by way of dissociating the uncanny from its misogynist overtones (the ultimate uncanny place for patriarchal culture is the mother's genitalia), it is necessary to make a distinction between the fascinatingly uncanny and the horrifically uncanny. Wetlands have by and large been the locus of the horrifically uncanny to be shunned, and destroyed; they need to become a place of the fascinatingly uncanny, their sights, sounds and smells, even their tastes and textures appreciated and conserved.

MODERN MUDFOG TO POSTMODERN WHITE NOISE

Such is the strength of the association between the wetland and the horrific in the modern European cultural unconscious that when Dickens would critically portray the Victorian industrial city with its evils and pollutions he represented it as an unhealthy marsh. In one of his little-known works, *The Mudfog Papers*, the title and eponymous town encapsulates the taxonomic anomaly of the wetland: mud as mixing earth and water, fog as a mixture of air and water. Water is thus a kind of promiscuous substance that gets around, even sleeps around, too much with earth and air, especially in that sink of iniquity the industrial town:

water is a perverse sort of element at the best of times, and in Mudfog it is particularly so. In winter, it comes oozing down the streets and tumbling over the fields, – nay, rushes into the very cellars and kitchens of the houses, with a lavish prodigality that might well be dispensed with; but in the hot summer weather it *will* dry up, and turn green: and although green is a very good colour in its way, especially in grass, still it certainly is not becoming to water; and it cannot be denied that the beauty of Mudfog is rather impaired, even by this trifling circumstance. Mudfog is a healthy place – very healthy; damp, perhaps, but none the worse for that. It's quite a mistake to suppose that damp is unwholesome: plants thrive best in damp situations, and why shouldn't men [*sic*]? . . . So, admitting Mudfog to be damp, we distinctly state that it is salubrious. The town of Mudfog is extremely picturesque. (Dickens (1880) 1987, p. 1)

The town is picturesque despite, or because of, in Dickens' ironical terms, its green water in summer. The irony upon irony is that it is a 'vulgar error', to use Dickens' term in a different context, certainly in the terms of twentieth-century scientific and medical discourse, though certainly a persistent one in popular culture, that damp causes disease.

'The health-preserving air of Mudfog' (ibid. p. 4) is precisely not health-preserving, not because of its dampness, but because of the pollutants being spewed out into its water and air by its factories to form the mudfog of its name. Like the fog figured as a yellow cat in T.S. Eliot's *The Waste Land*, the fog of Mudfog has a life of its own as it rises from the modern urban wasteland troped as wetland:

it had risen slowly and surely from the green and stagnant water with the first light of morning, until it reached a little above the lamp-post tops; and there it had stopped, with a sleepy, sluggish obstinacy, which bade defiance to the sun, who had got up very blood-shot about the eyes, as if he had been at a drinking party over night, and was doing his day's work with the worst possible grace. The thick damp mist hung over the town like a huge gauze curtain. All was dim and dismal. (Dickens (1880) 1987, p. 11)

In other words, all was like being in a wetland as it is conventionally represented and as Dickens evoked it in chapter 1 of *Great Expectations* where his narrator refers to 'the dark flat wilderness' of 'the marshes' which were, in turn, 'the dismal wilderness'.

The modern sunrise of Mudfog, unlike the postmodern sunset of Don De Lillo's 1984 novel *White Noise* which glows auratically and colourfully through the 'Airborne Toxic Event', is represented by Dickens as taking place in the miasmic atmosphere of a swamp, though the result seems to look more like the sun in a bush fire. The modern sun rises in the miasmic atmosphere of the mudfog coming from an English midlands factory town and the postmodern sun sets in the 'Airborne Toxic Event' of the noxious cloud above a middle American college town. These two events constitute the book ends (in at least two senses) of the rise and setting of the modern, the transition from high capitalism to late capitalism. The two books also mark a cultural and economic shift in emphasis from production to consumption, from the mudfog of modernity to the white noise of postmodernity. In this shift the black water of the wetland has become white noise, has become background interference and has been rendered largely invisible.

Even the titles of these two novels indicate a move from the modern technology of print on paper to the postmodern technology of electronic communication. With this shift comes a whole series of psychological and environmental changes. The modern sun is the black sun of melancholy and the postmodern sun is the black sun of depression, the postmodern analogue of or equivalent to melancholia. Yet the face of the black sun is obscured, not by the fog rising from the black water of a marsh, but by the air pollutants rising from the modern industrial capitalist city and factory. Postmodern industrial and electronic late capitalist technology produces, in turn, the media-simulated 'Airborne Toxic Event' whose precise ontological and scientific status no one can ever be sure about. The modern rural (picturesque) sky is reflected in still, canalised water; the modern urban (novelistic) sky is a miasmatic mudfog; the postmodern sky is white noise, 'the colour of a television tuned to a dead port', as in the famous opening line of William Gibson's classic cyberpunk novel *Neuromancer*, reflected in the black water of the wetland.

MODERN WASTELAND TO POSTMODERN WETLAND

First published one hundred years after the publication of Jewett's *A Marsh Island*, Graham Swift's *Waterland* continues the tradition of a novel set in a wetland which does not merely reproduce the pejorative (Swift 1984, p. 10). Here the English Fens are described as 'such a backward and trackless wilderness' (ibid. pp. 57, 58) and as a 'backwater' (ibid. pp. 13, 24 and 153), though not necessarily blackwater. *Waterland* is a typically postmodernist historiographical metafiction which neither fears nor valorises the seeming limitlessness of the Fen horizon with its 'flat, unimpeded outlook' (ibid. p. 152).[11] Crick, the first person narrator, off-handedly relates how 'the land in that part of the world is flat. Flat, with an unrelieved and monotonous flatness', its 'uniform levelness broken only by the furrowed and dead-straight lines of ditches and drains' (ibid. p. 2). Here, there are no pools of black water as in Haggard's *She*.

The land is flat because water has made it flat. Crick asks rhetorically: 'what is water ... which seeks to make all things level, which has no taste or colour of its own, but a liquid form of Nothing? And what are the Fens, which so imitate in their levelness the natural disposition of water, but a landscape which, of all landscapes, most approximates to Nothing?' (ibid. p. 11). The Fens and the wetland are the 'nothing landscape' (ibid. p. 44), or more specifically the nothing wetlandscape, the ultimate *terra nullius*. Even more precisely they are *aquaterra nullius*, the 'uninhabited' wetland to be colonised, the blank sheet, the *tabula rasa* to be inscribed by the dead(the operative word)-straight lines of ditches, drains, roads and railway lines.

The wetland can only be altered permanently by the technologies of road and railway line, canal and drain. The wetland walker can leave a mark with his or her feet on the blank sheet of the wetlandscape, but usually these are quickly erased and rendered back into the watery 'Nothing'. When William Byrd went on his commission in 1728 to survey the dividing line between Virginia and North Carolina through the Great Dismal Swamp one of the things he remarked repeatedly was the fact that 'the ground, if I may properly call it so, was so spongy that the prints of our feet were instantly filled with water' (Byrd 1966, p. 63). The wetland was not proper ground on which they could leave a mark. The 'men' were unable to make a lasting impression on the wetland as 'the impressions of the men's feet were immediately filled with water' (ibid. p. 70). Yet it made a lasting impression on them. And that is part of its horror, that it does not reciprocate in kind. Even when 'every step made a deep impression', these also were 'instantly filled with water' (ibid. p. 190). The soil was so spongy that 'the water oozed up into every footstep' the surveyors took (ibid. p. 191). The wetland provided 'no firm footing' (ibid. p. 70) on which the masculine and heroic surveyor could easily take a stand, legs astride, arms akimbo in order to see and master the land. Rather 'we found the ground moist and trembling under our feet like a quagmire' (ibid. p. 189).

The wetland is a dubious surface for walking upon, and surveying from. Yet rather than bemoaning the quaking surface of the wetland, other writers, such as Henry David Thoreau as we shall see in the final chapter, have affirmed it. 'The quaking zone' has even been seen as an alternative model of the self to the clean and proper, hard and dry, body and mind:

in one [patriarchal] aspect the self is an arrangement of organs, feelings and thought – a 'me' – surrounded by a hard body boundary: skin, clothes, and insular habits ... The alternative is a self as a centre of organisation, constantly drawing on and influencing the surroundings, whose skin and behaviour are soft zones containing the world rather than excluding it ... The epidermis of the skin is ecologically like a pond surface ... , not a shell so much as a delicate interpenetration. (Shepard 1971, pp. 24–5)

The self as a soft zone, as a swamp, rather than as a hard object, a castle, has 'the world' within rather than seeing it as something outside the self to master, not so much an interpenetration (delicate or not) as an intervagination. In patriarchal cultures swamps have by and large been reduced to surface and then penetrated by phallic heroes who *in*vaginate the surface, draw it back within itself as virgin (wet)lands in order to deny the depths of the mother.

Generally the wet land is not solid land. In the words of Liam Davison's *Soundings*, (a kind of Australian *Waterland*) 'what looked like solid ground could give way to water,' often abruptly and without warning (Davison 1993, p. 6). Graham Swift's *Waterland* remarks that 'the chief fact about the Fens is that they are reclaimed land, land that was once water, and which, even today, is not quite solid' (Swift 1984, p. 7). In the patriarchal western tradition, 'solid ground' as *Waterland* indicates (ibid. pp. 74 and 294) on which to plant ones feet, on which to build ones cities, in which to plants ones crops, from which to survey the surrounding plains and take in the view, has been valorised, if not fetishised.

Because of its lack of solidity and fluidity, the wetland poses specific problems for the builder of roads and boats. Western technology has largely gone hand in hand with western taxonomy when it comes to transportation across or through a wetland: as the latter has been divided been land and water, so has the former. Wetlands pose a problem for transportation: they are suitable neither for a deep-draughted boat nor for wheeled vehicles. The problem of transportation has also contributed to wetlands' poor showing aesthetically. The narrator of C. S. Forester's *African Queen* remarks of the Swamp of the Bora that for Allnutt and Rose 'no place could be beautiful that presented navigational difficulties' (Forester (1935) 1956, p. 135). Aesthetics is 'properly' subsumed beneath transportation. The aesthetic and the transportational are here posed as mutually exclusive, or more precisely hierarchical with the latter privileged over the former.

Instead of trying to find a suitable mode of transportation for the wetland, the western response has been largely to attempt to cross the marsh by building roads across or around it, or to canalise it in order to provide a means of communication through it. In cataloguing 'the more important physical factors modifying the formula of the Road' Hilaire Belloc in the 1920s mentions marshes first and foremost and then goes on to suggest that:

it is not always appreciated that the chief obstacle to travel from the beginning of time has been and still remains marsh, which may be defined as soil too sodden for travel, as distinguished from the lands which are boggy in wet weather but passable. Marsh is less striking to the eye, especially to the modern eye, than a stretch of water, much less striking than the apparent obstacle of the sea, or of a bold hill range: it is nevertheless the chief problem presented to the making of a road, because of all natural obstacles *it is the only one wholly untraversable by unaided man* [Belloc's emphases]. Man [*sic*] unaided can climb hills, swim water, work his way through dense undergrowth. But marsh is *impassable* to him: it is *the great original obstacle to progress* [my emphases]. (Belloc 1923, pp. 13–14)

Belloc shifts quickly from marsh as obstacle to travel to marsh as offensive to the eye, from engineering to aesthetics, both of which are nevertheless intimately associated in modernity, albeit in a merely functional hierarchical way as we have just seen in *The African Queen.*

The response of 'Progress' to the obstacle of the marsh has been to avoid it wherever possible ('our modern ways avoid marsh' (Belloc 1923, p. 15)), or to fill it or drain it where avoidance has not been possible, or more precisely where modern technology has made avoidance unnecessary, as in the case of cities such as St Petersburg or Perth (Western Australia). Indeed, if the wetland is the chief obstacle to progress as Belloc contends, then drainage is the epitome of 'Progress' (Davison 1993, pp. 103 and 115). In the process the land is reduced to surface to be travelled over or built upon with any impediment to either being constituted as an obstacle, not only to travel or cities, but to so-called progress, that motto of capitalism. The modern city, however, is characterised by not being compelled to avoid marshes. Modern technology, the engineering of dredgers, pumps, shovels and canals, made it possible for cities to go where no city had ever gone before, that is, into wetlands. Modern ways and the marsh are totally and irretrievably inimical to each other. They cannot co-exist; the marsh must go to make way for the modern.

Conversely, the return of the marsh for Belloc constitutes a return to the barbarian: 'whenever civilisation breaks down you begin to get a series of marshes, with all their accompaniments of fever and the rest growing up along the roads' (Belloc 1923, p. 65). Roads and drains are the bastions of modern industrial and urban civilisation holding out against the incursions of 'primeval marsh' (ibid. p. 104). Indeed, Belloc maintains that 'the main cause of the breaking down of the Roman Road was marsh' (ibid. p. 158). Could wetlands (the Pontine Marshes, the marshes of the Teutoburger Wald (Schama 1995, pp. 88–91), the Fens and other English marshes) even be a cause for the breaking down of the Roman Empire? Or any empire for that matter? Paddy fields, the Mekong Delta and the Plain of Reeds certainly helped to defeat American imperialism in Vietnam. Wetlands defy empire. Some wetlands, such as the Okefenokee Swamp in Georgia, have successfully 'resisted all attempts of [modern] human technology [or perhaps more precisely of American natural imperialism] to subdue and exploit it' (Bell Jr 1981, p. xi).[12] The marsh is also a metaphorical, even metaphysical place where meanings slip and slide, where morals go under, and reason cannot maintain its rectitude. Inimical to empire, empire must needs conquer it, or sink into the slimy depths, choke, suffocate and expire.

The swamp, and the wetland more generally, is even a smothering place, or perhaps more precisely a (s)mothering place, where various desires and fears about the mother's body are played out. In what David Miller calls 'the unfamiliar, smothering environment of the swamp, the ego is forced to face its long-repressed primitive [sic] urges' (Miller 1989, p. 24), perhaps a thinly-veiled allusion to Oedipal, even 'pre-Oedipal' desires. The swamp is both mothering of life on earth and smothering for the patriarchal male who would repudiate the indebtedness of his life to his mother's body and the dependency of his life on mother earth.

Yet rather than a place to bemoan the decline of empire, to bewail the decay of civilisation as we know it and to fear that the ramparts of civilisation will be swamped by the re-incursion of the barbarian, the wetland is a place to celebrate the ambivalence and fluidity of the postmodern. The wetland is the polysemic postmodern place par excellence, the place of multiple meanings where sense slips and slides never fixing on a point or a definition. Wetlands are the postmodern place par excellence partly because they are the pre-Oedipal place par excellence where the psychological and spatial repressed of the modern returns. Wetlands bear out Lyotard's contention that 'postmodernism . . . is not modernism at its end but in the nascent state, and this state is constant . . . *Post modern* would have to be understood according to the paradox of the future (*post*) anterior (*modo*)' (Lyotard 1984, pp. 79 and 81). Wetlands are the beginning state of modernism, the pre-modern wasteland out of which modernity arose and from which it was born, which modern technologies drained, or filled, or polluted, and which modernism celebrated or bemoaned or both ambivalently (e.g. Eliot's *The Waste Land*). Wetlands are the postmodern future anterior of the return of the repressed. They are the place of the remembrance of things past yet to come. They are the site of the anamnesia of the pre-Oedipal, the maternal, the uterine, the uncanny. Postmodern wetlands hark back to the pre-modern. Wetlands are the constant state of the birth of the modern.

THE DEAD-STRAIGHT LINE

The empire has made significant and irreversible inroads into reshaping the contours of the wetland, particularly in rendering its fluid shapes into the straight lines of drains, dykes, railways, roads, housing estates and canalised rivers. The pre-modern wilderness of the wetland having no economic value whatsoever is made over into the modern wasteland criss-crossed, or grid-ironed in Aldo Leopold's (1949. p. 100) apposite word, by drains, dykes, etc. H. C. Darby, the British historical geographer, relates how 'a bird's-eye view' of the Fenland in the Middle Ages:

would have revealed a country-side ranging in character from open pastures and meadows through reedy swamps to the pools of many meres connected by a confused network of channels. In dry summers, when the edges of the marsh dried up, the extent of the pasture was greater; while in winter, water might cover almost the entire face of the country. Around the islands, reclamation for cattle and for tillage was gaining upon the marsh; but, in the main, the effects of this reclamation were comparatively unimportant on the peat lands. The prospect today is very different. The airman [*sic*] looks down upon a regular pattern of channels separating well-tilled fields. Rivers run directly to the sea through corn, fruit, potatoes and sugar-beet. Straight lines dominate the scene. Small dykes divide the fields with mathematical regularity, and there are few hedgerows [those vestigial habitats of indigenous flora and fauna]. Long straight roads are frequent, and they cross each other at right angles. The railway lines are also characterised by long straight stretches . . . houses have dispersed themselves over the fen as well . . . in straight lines . . . (Darby 1956, p. 117)

The historical changes are significant: from the crooked lines and patchwork of pasture, meadows, meres and swamps to the straight lines and rectilinear grids of road, railway, fence, dyke, and housing; from the medieval to the modern, from chaos to order, from the pre-capitalist to the capitalist, from culture and nature in mutually beneficial symbiosis to parasitic agriculture and mono-culture. These transitions are seen and 'mapped' historically and dynamically (not in the static two dimensions of the map) from the air, the position from which Australian Aboriginal representations of the landscape often take place as in the famous dot paintings of the western desert, rather than from the ground, the position from which Constable's 'landscapes' with their reflections take place. The former is able to trace narrative and to read the country diachronically as song and story whereas the latter is locked (an appropriate metaphor) into seeing the pastoral country synchronically as private property to reflect upon statically and own as dead object. The former shows a number of different, sequential events occurring simultaneously in the dynamic space of the picture whereas the latter depicts one event frozen in time, and in the static space of the picture.

FLOWING OR STAGNANT WATER?

Water is, as Karl Wittfogel remarks, 'the natural variable par excellence' which 'flows automatically to the lowest accessible point in its environment', and there it tends to sit when it comes to wetlands (Wittfogel 1957, p. 15). Wetlands are by and large lowlands, even nether-lands. Water is also the cultural variable par excellence because the concept of water is an abstraction from various waters, whether it be waters of the seas, oceans, rivers, aquifers, or wetlands. Water from the tap and in the swimming pool is abstracted from various 'concrete' and particular waters. As Kelly Kelleher sinks beneath the waters of the marshland trapped in The Senator's car in Joyce Carol Oates' *Black Water* she finds that 'this water was not water of the sort with which she was familiar, transparent, faintly blue, clear and delicious not that sort of water but an evil muck-water, thick, viscous, tasting of sewage, gasoline, oil' (Oates 1994, p. 97). The cultural variability of water is here played out around its colour, clarity and transparency (or lack of them) through the senses of sight as well as those more immediate senses, taste and touch. The clear blue water in the swimming pool or the transparent delicious water out of the tap is not the dead black waters of the modern wetwasteland polluted by modern cities and industrial capitalist technology.

Water is also culturally variable depending on whether it flows or not. The water of rivers flow whereas the waters of wetlands stagnate. The distinction, even opposition in John Ruskin's case, between flowing and stagnant water is one of the fundamental organising principles of the dominant western cultural construction of nature. Yet, in turn, stagnant water has come to be associated more with foulness, to use Ruskin's term and re-use his association, than with stillness in a selective labouring of the psycho-geographical. The imagery of stagnant water can even be used as a powerful political emblem as in Wordsworth's poem 'London, 1802' in which he implores Milton that 'England hath need of thee. she is a fen/Of stagnant waters'.[13]

Yet for Ruskin and Wordsworth it was more than just that cleanliness and flowing water, foulness and stagnant water were associated. It was also that a whole series of moral and physical qualities went with them. For Ruskin, 'foulness is essentially connected with dissolution and death' (Ruskin 1888, II, p. 76). In Ruskin's moral topography, David Miller argues, 'light, life and the Deity are embodied in flowing water; darkness, death, and the withdrawal of the divine presence are exemplified in stagnant water' (Miller 1989, p. 81; Ruskin 1888, II, p. 77). After all, for Ruskin as a good Victorian, cleanliness was next to godliness, and dirtiness next to devilishness. The wetland was a moral emblem that troped sin allegorically as swamp, as it had been in 'The Slough of Despond' of Bunyan's *Pilgrim's Progress* some two centuries earlier. Significantly for Ruskin 'the rock', the hard and dry thing, even the turd, could be regarded as living, whereas stagnant water, the soft and wet substance, even menstrual blood or the placenta could not. This association has profound implications for considering the difference in cultural constructions of wetlands since, for example, Australian indigenes while regarding rocks as living earth regard wetlands as living water (Noonuccal and Noonuccal 1988).

Notes

1. I am very grateful to Wendy Parkins for drawing my attention to Blackwater Lake.
2. For Luce Irigaray (1993, p. 10), 'the relation to the mother is the "dark continent" par excellence'.
3. I am grateful to Wendy Parkins for drawing my attention to this novel.
4. See, for example, Terry Eagleton's references to 'the mire of historical relativism' and to 'a bourgeois class sunk in the mire of immediacy' (Eagleton 1991, pp. 91 and 97).
5. The relationship between the sublime and the slimy is discussed in Chapter 2 following Sofoulis' lead.
6. See also pp. 136–47 for a reading of Constable's 'Six-Footers'.
7. I am very grateful to Jon Stratton for alerting me to the pertinence of this book.
8. Similarly, contemporaneously and contiguously George Grey described how 'we were sitting on a gently-rising ground, which sloped away gradually to a picturesque lake, surrounded by wooded hills' (Grey 1841, p. 297).
9. I am grateful to Brian Dibble for drawing my attention to this poem.
10. See Thomas 1976; and Hochbaum (1944) repr. 1981. Hochbaum's drawings also illustrate Errington 1957.
11. For *Waterland* as a postmodernist historiographical metafiction, see Hutcheon 1988.
12. For American natural imperialism, see Kiernan 1978, especially the first part, 'The Winning of the National Territory'.
13. I am grateful to Margaret McIntyre for drawing my attention to this poem.

Philosophy in the Wetlands: The S(ub)lime and the Uncanny

The concept of the sublime is central to western aesthetics, especially in the modern period. Indeed, modern aesthetics for Jean-François Lyotard is 'an aesthetics of the sublime' (Lyotard 1984, p. 81).[1] Yet the concept is not of merely aesthetic interest and has a wider cultural pertinence to modernity in general and to the positioning of wetlands in particular. Around the name of the sublime, Lyotard remarked later, 'modernity triumphed', (Lyotard 1989, p. 199) not least over wetlands. Given the centrality of the sublime to the project of modernity, it is hardly surprising to recall that the concept-metaphor was important for both those colossi and critics of the modern, Marx and Freud. In the *Communist Manifesto* Marx and Engels used the metaphor of sublimation drawn from chemistry to argue that in bourgeois capitalism 'all that is solid melts into air' (see Figure 1). This phrase has since been taken up by Marshall Berman to encapsulate the etherealising project of modernity in which the solidities of pre-capitalist social formations evaporate into thin air (Berman 1983). By contrast, the obverse of sublimation, or desublimation, was used by what Klaus Theweleit calls fascist soldier males to represent their fear of revolution in which 'all that's solid becomes hot and fluid' (Theweleit 1987, p. 238), becomes, in other words, a tropical swamp, an object, or more precisely, abject of horror.[2] In opposition to fascism and to sublimation and desublimation (insofar as it is tied to its obverse) what Elizabeth Grosz calls 'a mode of transubstantiation, a conversion from solid to liquid', has been used by her as a feminist figure for orgasm (male and female) (Grosz 1995, p. 203).

A PSYCHOGEOCORPOGRAPHY OF MODERNITY
© Rod Giblett, 1994.

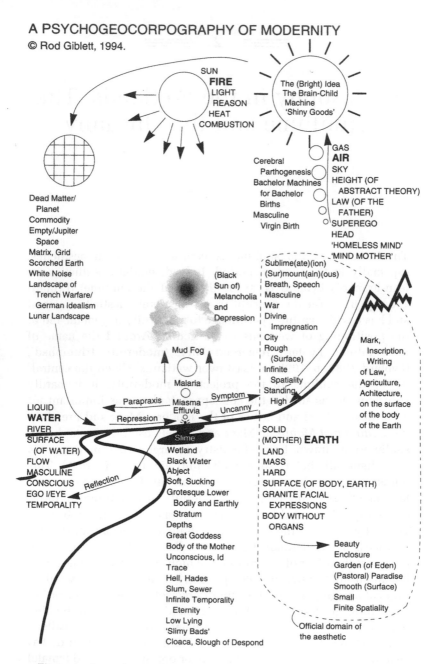

Figure 1   A psychogeocorpography of modernity (a drawing/mapping/ writing of/on the mind/body/earth in/by modernity).

Freud also used the metaphor drawn from chemistry to describe the process of sublimation by which sexual desire is displaced or deflected into ostensibly non-sexual realms, particularly the aesthetic and the intellectual (Freud 1985, pp. 39, 41 and 45; Laplanche and Pontalis 1973, pp. 431–4). The sublime and sublimation are closely linked as Kant attests that the sublimation of 'subduing one's passions through principles is sublime' (Kant 1960, p. 57).[3] The sublimation of sexuality into the aesthetic and intellectual can be seen as part of the condition of modernity, especially in the light of Simmel's and Spengler's argument that the intellect developed in conjunction with the rise of the modern metropolis. Indeed, for Spengler, 'the city is intellect' (Simmel 1950, pp. 409–24; Spengler 1932, pp. 92 and 96).[4]

Yet, despite, or perhaps because of the triumphs of modernity, sublimation is always, as it were, haunted by its shadow, its 'other'. The sublime, Zoë Sofoulis suggests, is always shadowed by what Freud calls the uncanny, the *unheimlich* or unhomely (Freud 1985, pp. 335–76). Rather than the reverse of the sublime, for Sofoulis 'the uncanny is the obverse of the sublime, its other side: that from which it springs and that into which its turns', and even that into which it returns (see Figure 1) (Sofoulis 1988, p. 12).[5] Sofoulis goes on to suggest that the uncanny is associated with slime, and that 'slime is the secret of the sublime', which she encapsulates in the parenthetical portmanteau 's(ub)lime' (ibid. p. 15, n. 86). The home of the slimy, and the uncanny, is the wetland summed up recently in the reference to 'the slime of the swamplands' (Tremayne 1985, p. 150).

In the Slime with Dante and on the Sublime with Longinus

Although the slimy achieved a certain sort of modern philosophical cachet by virtue of receiving a ten-page treatment in Sartre's *Being and Nothingness* (to which I return below), it is not a category or experience which has received much critical or theoretical attention despite its importance in *The Divine Comedy, Paradise Lost, Teenage Mutant Ninja Turtles* and *Ghostbusters I* and *II* amongst other texts.[6] In the seventh Canto (lines 103–30) and the fifth Circle of Dante's *Inferno* (and not the tenth level, as

the Mayor of New York surmises incorrectly in *Ghostbusters II*)
the condemned souls of the sullen, by a sort of sibilant (con-
sonantal), alliterative (patriarchal) justice (at least in John Ciardi's
English translation), are stuck in the slime of the Styx, transformed from river to marsh by Dante, complete with black water:

> By that foul water, black from its very source,
>   we found a nightmare path among the rocks
>   and followed the dark stream along its course.
> Beyond its rocky race and wild descent
>   the river floods and forms a marsh called Styx,
>   a dreary swampland, vaporous and malignant.
> And I, intent on all our passage touched,
>   made out a swarm of spirits in that bog
>   savage with anger, naked, slime-besmutched.
> They thumped at one another in that slime
>   with hands and feet, and they butted, and they bit
>   as if each would tear the other limb from limb.
> And my kind Sage: 'My son, behold the souls
>   of those who lived in wrath. And do you see
>   the broken surfaces of those water-holes
> On every hand, boiling as if in pain?
>   There are souls beneath that water. Fixed in slime
>   they speak their piece, end it and start again:
> 'Sullen were we in the air made sweet by the Sun;
>   in the glory of his shining our hearts poured
>   a bitter smoke. Sullen were we begun;
> Sullen we lie forever in this ditch.'
>   This litany they gargle in their throats
>   as if they sang, but lacked words and pitch.'
> Then circling on along that filthy wallow,
>   we picked our way between the bank and fen,
>   keeping our eyes on those foul souls that swallow
> The slime of Hell.

The slime of Hell is the product of the marsh or swamp of Hell,
and the obverse of the sublime of Longinus.

   If slime is the secret or obverse of the sublime as Sofoulis
contends, then the kind of qualities which Dante associates with
the sullen stuck in the slime would have their obverse in the
sublime. The earliest extant writing on the subject of the sublime in the western tradition is that of the treatise 'On the
Sublime' attributed to Longinus (Longinus 1965, pp. 99–158). In it

Longinus enumerates a number of qualities of the sublime which can quite readily be seen to have their obverse in the qualities Dante associates with slime. Whereas the sublime for Longinus exerts 'an irresistible force and mastery' (ibid. p. 100) (and thus can be considered masculine), slime for Dante exudes malignant vapours (and is considered feminine by Sartre); whereas the sublime for Longinus 'scatters everything before it like a thunderbolt' (ibid. p. 100), slime for Dante engulfs everything in it like a nightmare; and whereas the sublime reveals in a flash 'the full power of a speaker' (ibid. p. 100), slime stops up in a mouthful the power of speech.

Moreover, whereas the sublime 'uplifts our souls' (ibid. p. 107), slime is the damnation of the sullen; whereas the sublime fills one with 'a proud exaltation' (ibid. p. 107), the souls of the sullen expel a base exhalation in slime; whereas the sublime is associated with a sense of 'vaunting joy' (ibid. p. 109), slime is associated with degrading wrath; whereas in the sublime 'a noble emotion forces its way to the surface in a gust of frenzy' (ibid. p. 109), in slime an ignoble emotion forces its way to the surface in an expulsion of gas; and whereas the sublime breathes 'a kind of divine inspiration into the speaker's words' (ibid. p. 109), the sullen in slime exhale a kind of devilish incantation in broken words.

In addition, whereas the sublime involves 'the production of grand ideas, perpetually impregnating them with a noble inspiration' (ibid. p. 109) (the sublime is a metaphorical birthing machine (or a mechanical birth metaphor) for inseminating and giving birth to bright ideas, in short what has been called a Bachelor Machine for a Bachelor Birth[7]), the sullen in slime produce base gutturals, perpetually aborting them with an ignoble afflatus; whereas the sublime involves catching fire from the inspiration of others (Longinus 1965, p. 119), the sullen eat each other and the slime of hell; whereas the sublime impregnates one with heavenly power (ibid. p. 119), the sullen in slime are inflated with a hellish cannibalism; whereas the sublime inspires one to speak oracles (ibid. p. 119), the sullen in slime are inflated to emit noise; and finally, whereas the sublime carries one up close to the majestic mind of God (ibid. p. 147), the sullen in slime are carried away close to the debased lusts of the Devil.

ON THE SUBLIME AND BEAUTIFUL WITH KANT AND BURKE

The concept-metaphor of the sublime is later taken up by Burke and Kant who rework it, especially in relation to the sensations with which it is associated, though it is still inevitably haunted by its slimy 'other'. For Burke the sublime is a delight which turns on pain. It also causes astonishment, 'that state of the soul in which all its motions are suspended with some degree of horror' (Burke 1958, p. 57). For Kant, though, the astonishment of the sublime amounts almost to terror (Kant 1952, p. 120 and 1960, p. 48). By contrast, the uncanny for Freud is related to what is frightening and to what arouses dread and horror (Freud 1985, p. 339). Whereas the sublime is attended with 'some degree' of horror, the uncanny arouses exclusively dread and horror. The ruling principle of the sublime for Burke is terror (Burke 1958, p. 58), whereas the ruling principle of the uncanny for Freud is horror.

Terror and the sublime, pleasure and the beautiful, could be said to be the ruling aesthetic responses and cultural passions of modernity whereas horror and the uncanny may be those of postmodernity. The sublime implies an impossible object[8] (such as dead rock) terrifying in its magnitude whereas the uncanny (and slimy) intimates an impossible abject (such as living water) horrifying in its monstrosity and the beautiful implies a possible object pleasing in its finitude (such as a gentleman's park or a feminine woman). The first is arguably associated with the exteriority of the Law of the Father, whereas the second is linked with the interior of the Body of the Mother and the third with the shape and surface of the body of the idealised woman in patriarchy. The distinction between the beautiful and the sublime has been gendered with the former being ascribed to the female/feminine and the latter to the male/masculine (Kant 1960, pp. 77 and 93).

Yet the sublime, terror, horror and the uncanny are not just mere aesthetic responses but modalities which set up and involve relations between subjects and objects, or abjects (neither subject nor object but in between) in the case of the uncanny. The sublime is a feeling of awe and fear in the face of gigantic objects, such as mountains, insofar as they are representatives for the ultimate symbolic object, the Phallus and the Law of the Father.[9] The uncanny is a feeling of awe and fear in response to imaginary abjects, such as smells, insofar as they are images of the ultimate imaginary abject, the Body of the Mother. Finally, horror is an emotion comprising fear and loathing in response to more immediate, but feared, contact with the slimy and the swampy insofar as they evoke memories of the Body of the Mother.

The beautiful, though, is quite different. For Burke, the sublime dwells on great and terrible objects, whereas the beautiful dwells on small and pleasing ones (Burke 1958, p. 113). Similarly for Kant 'the sublime must always be great; the beautiful can also be small. The sublime must be simple; the beautiful can be adorned and ornamented' (Kant 1960, p. 48). Whereas the sublime has, or is associated with, a broken and rugged surface, the beautiful for Burke has, or is associated with, a smooth and polished surface (Burke 1958, p. 114). Whereas the sublime concerns or represents vast objects, the beautiful, comparatively small objects. Whereas the sublime is dark and gloomy, the beautiful is not obscure (ibid. p. 124). Whereas the sublime is founded on pain, the beautiful is founded on pleasure and acts by relaxing the solids of the whole system and so is the cause of all positive pleasure (ibid. pp. 149–50). The beautiful is a kind of pleasurable enemetic evacuation or spermatic ejaculation, whereas the sublime could be characterised as the pleasure/pain of 'a good shit' of solid objects, or turds. The sublime entails expelling and transcending objects and matter. Thus Deleuze and Guattari have remarked that 'sublimation is profoundly linked to anality' (Deleuze and Guattari 1977, p. 143).

On the Uncanny and Sublimation with Freud

The distinction between the sublime and the beautiful has been clearly and traditionally marked but what of the uncanny and the sublime? For Burke the sublime is associated with the dark and gloomy (Burke 1958, p. 124) and to some extent for Freud the uncanny is likewise associated with the dark and difficult to see. This is particularly the case in the famous story that Freud relates from a magazine he happened to read:

I read a story about a young married couple who move into a furnished house in which there is a curiously shaped table with carvings of crocodiles on it. Towards evening an intolerable and very specific smell begins to pervade the house; they stumble over something in the dark; they seem to see a vague form gliding over the stairs – in short we are given to understand that the presence of the table causes ghostly crocodiles to haunt the place, or that the wooden monsters come to life in the dark, or something of that sort. (Freud 1985, p. 367)

Unlike the sublime, which for Kant was associated with the formless and the devoid of form, and the beautiful which he associated with form (Kant 1952, p. 90), the uncanny for Freud is associated with a vague form: it has neither the formlessness of the sublime nor the form of the beautiful. In this respect, then, the uncanny is thus not so much the obverse of the sublime as in-between, or the mediating category between, the sublime and the beautiful.

Freud's account of the uncanny is crucial for a number of other points. In the story above the uncanny is produced by the smell of virtual crocodiles, the kings of the tropical wetland, which is the obverse of the temperate dry land. The crocodile is the archetypal swamp monster. Furthermore, for Ruskin, it is 'slime-begotten'[10] though, as Camille Paglia puts it, slime is also for humans 'our site of biologic origins' (Paglia 1991, p. 11).[11] The story Freud retells is also significant because of its emphasis on the sense of smell. It is no surprise that Hoffmann, who was for Freud master of the uncanny should, in his story 'The Sandman', in which Freud found the uncanny most powerfully evoked, refer to 'a subtle, strange-smelling vapour . . . spreading through the house' (Hoffmann 1982, p. 88). For Freud the uncanny is *unheimlich*, unhomely, but also homely, contradictory feelings which Freud found associated with the first home of the womb in the minds of adult males.

In both stories the unhomely invades and pervades the homely via a distinctive and unpleasant smell. The uncanny is mainly a matter of smell whereas the sublime for Burke is primarily a matter of sight (Burke 1958, p. 57), as is sexual difference in patriarchy. The sense of sight was privileged in and by aesthetics whereas the sense of smell was disqualified from it by Kant and later Hegel (Corbin 1994, pp. 7 and 229). Eagleton has argued that 'only certain sensations are fit for aesthetic enquiry; and this means for the Hegel of the *Philosophy of Fine Art* only sight and hearing . . . There can be no aesthetics of odour, texture or flavour' (Eagleton 1990, p. 33). Of course, there are pleasant and unpleasant smells, touches and tastes, but they are not formally aestheticised.

Hence the lack of an aesthetics appropriate to the smells, textures and tastes of the wetland. As the wetland does not conform to the dictates of form, shape and vista of the beautiful, the sublime and the picturesque, it is has been beyond the pale of aesthetics on every count. The sensations of sight and hearing have been limited to the beautiful, sublime and picturesque. The sights and sounds, not to mention the smells, tastes and textures, of the wetland have been systematically excluded from aesthetics.

The uncanny as an aesthetics of smell is a kind of anti- or counter-aesthetics, especially to the sublime. It disrupts phallocentric sexual difference and its privileging of sight since the smell of the uncanny can be and has been strongly associated with female sexuality (Gallop 1982, p. 27). The uncanny is associated with smell but not just with any smell. The smell which evokes the uncanny is what Freud calls 'a very specific smell' which emanates from the virtual crocodiles but is not closely associated with them. Freud does not name this smell hence represses it as he does generally with the sense of smell relegating references to it to what Gallop calls his 'smelly footnotes' (ibid. p. 28). But it could be associated, given the lack of a discriminating aesthetics of smell, in the patriarchal mind, or to its nose, with the odour of female genitalia, which for Freud were the ultimate *unheimlich*, the (un)homely or uncanny.

This uncanny odour is, however, not the passive reverse of the sublime sight, but its active obverse, a kind of anti- or de-sublimation. The odour of female genitalia, the *'odour di femina'*, Gallop argues, 'becomes odious, nauseous, because it threatens to undo the achievements of repression and sublimation, threatens to return the subject to the powerlessness, intensity and anxiety of an immediate, unmediated connection with the body of the mother', and with 'Mother Earth', I would add, and with those regions, such as swamps, with similarly offensive odours to the non-discriminating patriarchal nose (Gallop 1982, p. 27).

The uncanny smell of the wetland entails a return to the repressed, as do uncanny smells in general (see Figure 1). This return is associated by Freud with the artefacts of colonialism which bear the traces of other, alien or exotic places and peoples. Freud construes this return to the repressed as leading back to an old, animistic conception of the universe (Freud 1985, p. 363). These animistic stages leave traces which, for Freud, manifest themselves and are expressed in the uncanny. In other words, in the uncanny the traces left by the animistic stages, or more precisely in this context, left by destruction of wetlands, are retraced.

No space, Henri Lefebvre argues, 'ever vanishes utterly, leaving no traces' (Lefebvre 1991, p. 164). Filled or drained wetlands have not vanished entirely leaving no traces. For Lefebvre, like Freud, 'the past leaves its traces', yet for Lefebvre, unlike Freud, places also leave their traces (ibid. p. 37). The uncanny could be seen to be the obverse of the Freudian symptom. Unlike the symptom which, for Freud, leaves traces of the unconscious on the surface of the body of the patient, the uncanny reworks the traces and returns the subject momentarily to the unconscious and to the surfaces of the body (see Figure 1) which have been inscribed, and some of whose depths have been invested, by capitalism (Freud 1973, pp. 296–312).

The process of sublimation can be seen, then, to be an extension of the symptom in that it moves the subject even further away from contact with the unconscious (see Figure 1). Sublimation has recently and critically been reconstrued by Klaus Theweleit as 'a form of male dominance over the earth, women, and men's own affects' (Theweleit 1987, p. 238). This dominance can be articulated around the metaphorics of different regions of the earth and even different activities on the earth, such as agriculture, as they are used to figure women's bodies and men's affects, and men's bodily, physical life. This history, this his-story, has been traced by J. J. Bachofen, the nineteenth-century ethnologist and Hegelian jurist, who relates how:

the triumph of paternity brings with it the liberation of the spirit from the manifestations of nature, a sublimation of human existence over the laws of material life . . . Spiritual life rises over corporeal existence, and the relation with the lower spheres of existence is restricted to the physical aspect. Maternity pertains to the physical side of man, the only thing he shares with the animals: the paternal-spiritual principle belongs to him alone. (Bachofen 1967, p. 109)[12]

For Kant, similarly, the sublime, or at least what he calls the dynamically sublime, which he defines as 'nature considered in an aesthetic judgement as might that has no dominion over us' (Kant 1952, p. 109), 'reveals a faculty of estimating ourselves as independent of nature, and discovers a pre-eminence above nature' (ibid. p. 111) in the realms of the transcendental and the extraterrestrial. Indeed, for Kant, the dynamically sublime is the means by which we 'become conscious of our superiority over nature within and thus also over nature without us' (ibid. p. 114), over the immanent (including the (aqua)corporeal) and the terrestrial.

By contrast, the mathematically sublime for Kant is the ideas of reason considered in a reflective judgement as might that has dominion over us. It reveals a faculty of estimating ourselves as inferior to the ideas of reason and discovers a pre-eminence above 'us': 'it represents our imagination in all its boundlessness, and with it nature, as sinking into significance before the ideas of reason' (ibid. p. 105). The mathematically sublime reveals a faculty for considering not only 'ourselves' but also nature as inferior to the ideas of reason in the realms of the ideational and the rational above, as it were, the transcendental and extra-terrestrial.

The 'double mode of representing an object as sublime' (ibid. p. 94), as Kant called the mathematically and dynamically sub-lime, sets up a hierarchical triangulation with the ideas of reason as the privileged term and nature as the denigrated term with 'ourselves' as the mediating term. The mathematically sublime sets up the superiority of the ideas of reason over nature and 'ourselves', and the dynamically sublime the superiority of our-selves over nature. On the one hand, 'we' can feel independent of and superior to nature within and without, and on the other inferior to the ideas of reason along with nature. 'We' and nature are put in our place vis-a-vis the ideas of reason and 'we' put nature in its place vis-a-vis 'ourselves' and the ideas of reason. The philosophy of the (mathematically and dynamically) sublime and the psychology of sublimation underpins and enables idealist, rationalist mastery of nature, not least of the aquaterrestrial.

Bachofen historicised this process of sublimation more extensively in terms of stages of human development, crucially in the transition from what he calls the lower Aphroditean-hetaeric beginnings 'born in the lustful chaos of primeval swamps' as Deleuze puts it, through the higher Demetrian-conjugal stage to the Apollonian-patriarchal pinnacle (Deleuze 1989, pp. 52 and 53). For Bachofen, 'the former corresponds to the unregulated swamp generation, the latter [two] to ordered agriculture' (Bachofen 1967, p. 195). In terms of sublimation, the first stage for Bachofen is associated with the repudiated Body of the Mother which I have linked with horror and the last with the rigidified Law of the Father which I have tied to terror. He makes a 'contrast between agriculture and the *iniussu ultronen creatio* [the ultimate unusable creation], the unbidden wild growth of mother earth, manifested most abundantly and luxuriantly in the life of the swamps' (ibid. p. 97). 'The wild vegetation of the swamps' is for Bachofen 'the model of motherhood without marriage' (ibid. p. 113), the dreaded 'unmarried mother' who still evokes the same moral panic today as she is 'available' and desirable (because unmarried) and taboo (because she is a mother).

In the transition to agriculture a new type of motherhood arises. With the development of agriculture, Bachofen argues:

> motherhood takes on a new significance, a higher form. The wild swamp generation, which eternally rejuvenates matter in everlasting self-embrace, which brings forth only reeds and rushes or the 'swampy offspring of the sources' [Aristophanes, 'The Frogs'], and which springs up uselessly without regard to man, is replaced by the act of the tiller of the soil who opens the womb of the earth with his plow, who lays the seed in the furrow, and harvests nutritious fruit, Demeter's food. (ibid. p. 191)

The process of sublimation can be construed in psychoanalytical terms via the way in which individual development (or ontogenesis) repeats the development of the species (or phylogenesis) as a move out of the swamps of the unconscious into the tilled fields of the surface of the earth/body (see Figure 1).

The metaphor of the unconscious as swamp, David Miller has argued, emerged during the 1850s in order to trope what he calls 'newly awakened unconscious mental processes' (Miller 1989, p. 3) in modernity. The metaphor has recently been taken up and used in postmodernity to great effect:

the Marshland, a zone that is transparent and fetid at the same, a place that looks like a crystalline paradise and is in reality a swamp, formed of decomposing plants, bogs, poisonous flowers, and all manner of vermin. Could you find a better representation than this of the human unconscious? The clarities that conceal, the putrid smells that surface with each step. We accept it as a real place because that's the way we are in the darkest regions of our being. That's why I insist that the Marshland is a symbolic location. (Valenzuela 1987, p. 48)

Whereas sublimation is the process of transforming solid into gas, crystallisation is the processing of transforming liquid into solid, the most obvious everyday example being honey. The swamp looks solid, but is in fact liquid, or more precisely, it is between the liquid and the solid, one of, if not the, most disturbing of the qualities ascribed to the wetland in patriarchal western culture.

Sublimation links up with, or more precisely continues, the process of crystallisation to transform base matter into the ethereal monuments of capitalist modernisation, such as the Crystal Palace. The results of the process of sublimation are quite profound for the 'solid matter' and liquid flows of the human body. Once you have sublimated the body, Luce Irigaray remarks, 'all you have left is air, smoke, vapour, ghost . . . it is nothing but airs and phantoms' (Irigaray 1985, pp. 283 and 341). In the western tradition these airs and phantoms have been made to come from the wetland, whether it be the bad and rotten airs which 'cause' malaria in the miasmatic theory of disease or the phantoms of monsters and other fantastic creatures which haunt the swamp of horrors. Yet rather than seeing bad airs and swamp monsters as intrinsic features of the wetland, it would be more apposite to see them as products of sublimation (see Figure 1).

The process of sublimation finds its supreme manifestation in the (al)chemical process of taking the 'base or dead matter' of the swampy and the slimy and transforming it into the 'gaseous heights' of philosophy, abstract theory and the superego (and the 'gas' to power cars) (see Figure 1). Modern civilisation is fuelled by swamp power (Mitsch and Gosselink 1986, p. 3) both in its coal-powered phase of colonialist and expansionist 'progress' and in its petroleum-powered phase of neo-imperialist and managerialist 'development' (Adams in Gare 1995, pp. 5–6) which made possible the modern domination of the earth including, ironically, wetlands (McKibben 1990, p.137). Pre-modern Roman civilisation was also dependent upon the products of the swamp as Pliny the Elder argued in the first century CE that 'our civilisation – or at any rate our written records – depends especially on the use of paper' made from papyrus which 'grows in the swamps of Egypt and in the sluggish waters of the Nile where they have overflowed and form stagnant pools not more than three feet deep' (Pliny 1991, pp. 175–6). Empire is dependent on wetlands.

Yet the process of sublimation does not end with the completion of this etherealising and transcending project but continues with the denigration of originary matter, the earth and water, the slime of wetlands, much as Socrates undertook in the *Phaedo*:

there are many hollow places all around the earth, places of every shape and size, into which the water and mist and air have collected. But the earth itself is as pure as the starry heaven in which it lies, and which is called Ether by most of our authorities. The water, mist, and air are the dregs of Ether, and they are continually draining into the hollow places of the earth. We do not realise that we are living in its hollows, but assume that we are living on the earth's surface . . . this earth and its stones and all the regions in which we live are marred and corroded by the brine, and there is no vegetation worth mentioning, and scarcely any degree of perfect formation, but only caverns and sand and measureless mud, and tracts of slime wherever there is earth as well; and nothing is in the least worthy to be judged beautiful by our standards. But the things above excel those of our world to a degree far greater still. (Plato, *Phaedo* 109A–110D)

Socrates valorises the extraterrestrial and the transcendental over the immanent – the intraterrestrial and the aquaterrestrial. The western theory of the s(ub)lime could be seen to have its formalised beginnings here not least in its enunciation of the Platonic theory of forms and of the sublime as a faculty, as it was later for Kant, for estimating ourselves as independent of and superior to nature, including the mother's body, as a defence against the power of nature.

In the Slime with Sartre

Although, as I remarked earlier, the slimy is a crucial category in *The Divine Comedy, Paradise Lost, Being and Nothingness, Teenage Mutant Ninja Turtles* and *Ghostbusters I* and *II*, it has not received much critical and theoretical attention. It is, however, a crucial category and figure in the western cultural tradition, not least, as Sofoulis argues, because it is the secret of the sublime. Like the sublime, slime is an in-between, mediating category. The swamp also, as David Miller has remarked, is 'neither land nor water' and 'represents an in-between state' (Miller 1989, p. 175). (see Figure 1) Such a swamp is the Swamp of the Bora in C. S. Forester's *The African Queen* which is described as 'a dreary, marshy, amphibious country, half black mud and half water' (Forester (1935) 1956, p. 106). The tropical swamp is half land and half water, including its bottom, if it could be said to have one. The bottom of the swamp was 'deep semi-liquid mud' (ibid. p. 124), and the mud itself is described as 'half water, as black and nearly as liquid' (ibid. p. 136). Whereas the sublime is in-between the solid and the gaseous, slime is between the solid and the liquid, and the swamp also is between the land and the water. It is also between the light and the dark. Allnutt and Rose 'never saw the sun while they were in that twilight nightmare land' (ibid. p. 139). The swamp is, in Miller's words, 'a dangerous realm where all distinctions blur' (Miller 1989, p. 23).

Besides their geo-politics, the sublime, the slime and the swamp have a gender politics: whereas the sublime has been construed as masculine, slime has been construed as feminine, not least by Sartre, as the swamp has also been feminised and the female body aquaterrised, for example by Camille Paglia in an essentialist, ahistorical gesture: 'bog and quagmire are my chthonian swamp, that dank primal brew of earth and water that I identify [though she is hardly the first] with the female body', or with certain regions of it (Paglia, p. 92; see also p. 591). Similarly, the 'grotesque lower bodily stratum' as Bakhtin (1984) coyly puts it of the 'dark continent' of female sexuality has been figured as swamp, for example, in Vladimir Nabokov's novel *Ada* in which Van Veen grabs from behind the 'hot little slew' of the eponymous central female character (Nabokov 1971, p. 98). 'Slew' is, according to the *Shorter Oxford English Dictionary*, 'a marshy or reedy pool, pond, small lake, backwater [b(l)ackwater?], or inlet.' A 'hot little slew' is, in other words, a minor invaginated, culturally devalued, swamp. A slew is also a term used in coal-mining to describe 'a natural swamp in a coal seam'. Yet if the wetland is going to be feminised, or if it is going to be troped as a part of female anatomy or an aspect of female physiology (or the female body aquaterrised), it should be in terms of the placenta: the placenta is the wetland of the maternal body; the wetland is the placenta of the Great Mother (Earth) (Irigaray 1993, pp. 38–44).[13]

Slime is construed as feminine in Sartre's *Being and Nothingness*. In about ten markedly misogynist pages of his *magnum opus*, Sartre undertakes a philosophical meditation on what he calls 'the horror of the slimy' (Sartre 1969, p. 611). For Sartre, the slimy is feminine and an object of horror for what he calls 'the European adult' (ibid. p. 604), though I would suggest he is referring to the white male adult. The white male adult is constituted as such by not being the black feminised water of the wetlands and by being horrified by the (s)mothering body of the swampy mother.

In Sartre's own philosophical terms, the slimy represents the revenge of the in-itself, or solid, exemplified in 'the immanent body', on the for-itself, or fluid idealised in 'consciousness and reflexive self-consciousness' (Grosz 1994, p. 86; see also Figure 1 above). Part of the horror of the slimy resides in the fact that it is neither solid, nor fluid, as part of the horror of the swampy lies in the fact that it is neither earth, nor water. The slimy and the swampy are neither one thing nor the other, but occupy an unsettling middle position between two things. They both fall into, or are more precisely, a mediating category between the two extremes of solid earth and liquid water (see Figure 1). Yet for Sartre the in-betweenness of the slimy does not represent a settled state or a stasis. Rather, the slimy represents for Sartre the solidifying of the fluid, the damming of its flow, the damning of the flows. For him 'a slimy substance is an aberrant fluid', a fluid which does not flow in the normal channels.

As slime is an aberrant fluid, it does not let itself be contained, and it does not allow one to make a mark on it, or more precisely, to make one's mark on it. For Sartre, the slimy is a becoming 'on which one leaves no mark, and which could not leave a mark on us' (Sartre 1969, p. 607). The slimy, like, or more precisely *as*, the swamp, does not allow the writing of development to take (a) place on the earth unless and until it is filled or drained. It does not allow unaltered any fixtures which are permanent. For Sartre, the slimy is 'a being which is eternity and infinite temporality' (ibid. p. 607). It does not permit the stasis of western 'man-made' structures.

The slimy, however, is ambiguous because it represents for Sartre, at least partially, 'a dawning triumph of the solid over the liquid' (ibid. p. 607), as the Swamp of the Bora in *The African Queen* 'was a region in which water put up a good fight against the land which was slowly invading it' (Forester (1935) 1956, p. 136), a fight which nevertheless it is doomed to lose. On the one hand, the slimy and the swamp obstruct the solid from taking (a) place, and on the other, the slimy, and the swamp, are the invasion and triumph of the solid earth over the liquid water. It is this undecidability of the slimy that makes it such an abject of horror. It is neither solid nor liquid, but it just does not sit still as some sort of fixed and static mediator. It is both the process of one becoming the other (the liquid solidifying), and the prevention of the one taking the place of, or taking place on, the other (the solid being erected on it). For Sartre, the slimy is: 'a tendency of the indifferent in-itself which is represented by the pure solid, to fix the liquidity, to absorb the for-itself which ought to dissolve it. Slime is the agony of water. It presents itself as a phenomenon in process of becoming' (Sartre 1969, p. 607). Hence the slimy is horrific because it is not being but becoming.

In its becoming, the slimy for Sartre represents what he calls 'a constant hysteresis in the phenomenon of being transmuted into itself' (ibid. p. 608). As Susan Griffin (1978, p. 13) has reminded us, the word 'hysteria', and its derivatives such as hysteresis, come from the Greek word 'hyster' meaning 'womb'. Sartre uses the word 'hysteresis' pejoratively to refer to matter being transmuted into itself in the womb, rather than into something other than itself; matter enwombed in slime to produce more matter rather than matter sublimated into the sublime to produce non-matter. Sartre's use of the word 'hysteresis' can also be read historically in Bachofen's terms as a patriarchal supersession of the matriarchal, of the swampy and slimy womb of life by the agricultural.

The becoming of the slimy contrasts strongly with the being of the solid whose inertia for Sartre 'symbolises for me my total power' (Sartre 1969, p. 608). The solid is prototypically the turd which is ideally not so soft and malleable to be slimy, nor so hard and dry to be intractable. The solid turd is also prototypically the first form of the gift from the child to the mother whereas the slimy placenta is the first gift from the mother to the foetus (Rouch in Irigaray 1993, pp. 43–4). The placenta is a more primary gift than the turd – the placenta is the first gift. The qualities of objects have implications for the power which can be wielded against or over them. Whereas the solid symbolises for Sartre his total power, presumably, by implication, the slipperiness of the slimy, its aberrance, symbolises for him a loss or defeat of mastery. As I have already remarked, the sublime is the pain/pleasure of a 'good shit', and the beautiful is a kind of enemetic evacuation, or spermatic ejaculation, whereas the slimy is a slipping away, perhaps associated with diarrhoea, but also a sucking in.

Power over the turd resides in being able to expel or eject it whereas the horror of the slimy consists in its being able to ingest. The slimy for Sartre, like the swamp in patriarchal western culture:

sticks to me, it draws me, it sucks at me. Its mode of being is neither the reassuring inertia of the solid nor the dynamism like that in water which is exhausted in fleeing from me. It is a soft, yielding action, a moist and feminine sucking, it lives obscurely under my fingers, and I sense it like a dizziness; it draws me to it as the bottom of a precipice might draw me. There is something like a tactile fascination in the slimy. I am no longer the master in arresting the process of appropriation. It continues. In one sense it is like the supreme docility of the possessed, the fidelity of a dog even when one does not want him any longer, and in another sense there is underneath this docility a surreptitious appropriation of the possessor by the possessed. (Sartre 1969, p. 609)

The slimy upsets the solidities of the positions of master and possessor, and indeed threatens to reverse the roles. Sartre's horror of slime stems from the fact that it jeopardises his attempts to achieve a transcendental identity since he was born from the slime of the mother and tries to separate himself off from it/her.

Sartre goes on to describe this possession by the possessed of the possessor as 'poisonous' (ibid. p. 609), whereas presumably by implication the possession by the possessor of his possessions, his solid objects, his turds, is not. Sartre then switches metaphor to talk about 'the snare of the slimy: it is a fluidity which holds and which compromises me, . . . it clings to me like a leech . . . and leaves its traces upon me' (ibid. p. 609). No one can leave a mark on the slimy as William Byrd found with the Great Dismal Swamp. The slimy, however does leave its traces as Sartre says. The swamp not only 'attacks the surfaces of the body' (Miller 1989, p. 178), but also marks them. Rather than themselves being a surface for inscription, the slimy and swampy mark the surfaces of the body. The slimy and the swampy not only (re)present an impossible writing surface, but are also collectively arche-writing or trace (see Figure 1). For Sartre, 'the slime is like a liquid seen in a nightmare . . . Slime is the revenge of the In-Itself. A sickly-sweet feminine revenge' (Sartre 1969, p. 609). The slimy leaves its traces across most of the senses – it tastes sickly-sweet, it looks liquid (or solid) but isn't really, it feels fascinating, but 'to touch the slimy is to risk being dissolved in sliminess' (ibid. p. 610).

Slime is not only problematic spatially but also temporally. Whereas for Sartre the slimy is associated with eternity and infinite temporality, following Bergson, the duration of consciousness is like a river (ibid. p. 610). As the slimy is construed as eternity and infinite temporality and marked as feminine, so for Sartre the river is duration of consciousness and marked as masculine (see Figure 1). In the light of this discussion there are additions to be made to the list of qualities which have been ascribed to the masculine and the feminine in patriarchal western culture–

Activity	Passivity
Culture	Nature
Father	Mother
Head	Heart
Man/Masculine	Woman/Feminine
Reason	Emotion
Action	Speech

Doing	Becoming
River	Swamp
Flow	Stagnation
Sublime	Slime
Duration of	Eternity and
consciousness and	infinite temporality[14]
infinite spatiality	

Sartre associates the slimy with holes, holes in the sand and in the earth (Sartre 1969, p. 612), though he does not mention wetlands. The function of a hole for Sartre is to be filled and this filling is construed in masculinist sexual terms. For him, 'the hole is originally presented as a nothingness 'to be filled with my own flesh' (ibid. p. 613). The sexual act is construed here in masculinist terms of filling holes, whether it be holes in human bodies (though which holes in whose body Sartre does not specify), or holes in mother earth. For Sartre: 'the ideal of the hole is then an excavation which can be carefully moulded about my own flesh in such a manner that by squeezing myself into and fitting myself tightly inside it, I shall contribute to making a fullness of being exist in the world' (ibid. p. 613). This desire for a snug hole does not differentiate between anus and vagina, and so includes the anal eroticism of penetrating and trying to fill the anus of mother earth.[15] The slimy and the swampy in matriarchy had been the womb of life whereas in patriarchy they have become the anus of the earth where wastes are excreted or dumped. Patriarchy then fails to differentiate between the two regimes and regions and treats the former as it does the latter. Yet Sartre's desire for a snug hole is also a recognition of lack and a desire for union and completion, if not for procreation.

Sartre triumphantly announces that 'here at its origin we grasp one of the most fundamental tendencies of human reality – the tendency to fill' (Sartre 1969, p. 613). If anything could be said to characterise the history of wetlands in patriarchal western culture it would be this tendency to fill, or more precisely – as 'tendency' is a euphemism if there ever were one – the will to fill.

With boyish good humour Sartre goes on to say that 'a good part of our life is spent in plugging up holes' like the Dutch boy sticking his finger in the dyke, 'in filling empty places, in realising and symbolically establishing a plenitude' (ibid. p. 613) which produces a full, replete and smooth surface and which destroys by filling. As slime or wetland is to the unconscious, so dykes and drainage are to sublimation. He blithely adds that 'it is only from this standpoint that we can pass on to sexuality' (ibid. p. 613) as if sexuality were not already present in his desire for a snug hole. What he seems to be suggesting is that it is only from the standpoint of the will to fill that he can pass on to a discussion of female sexuality, so implicitly admitting that he has been concerned with male or masculine sexuality as normative.

Whereas Freud associates the uncanny with female genitalia, Sartre associates them with the slimy. For both of them uncanny and slimy female genitalia are disturbing and disrupting. For Sartre, 'the obscenity of the feminine sex is that of everything which 'gapes open'. It is *an appeal to* being as all holes are' (ibid. p. 613). In other words, it is not being but a lack of being. Yet is not equally Sartre's desire for union an appeal to being? For Sartre it seems, as for the fascist soldier male, and as Barbara Ehrenreich puts it 'women's bodies are the holes, swamps, pits of muck that can engulf' (Ehrenreich 1987, p. xiii). But women's bodies could only be seen as such by confusing anus and vagina/womb (the excretory and reproductive/sexual functions) and so denying sexual difference (with women because men have an anus too) (Kristeva 1986, p. 147) and simultaneously polarising gender in non-reproductive terms (across men and women). By desiring indiscriminately a snug hole, and so failing to differentiate between vagina/womb as source of life and anus as site of waste, Sartre denies sexual difference with women, especially women's capacity to procreate.

At the same time, by associating women's bodies with slime and holes, and by associating men's bodies with solid objects, Sartre polarises gender in non-reproductive terms across men and women. This is, as Jessica Benjamin and Anson Rabinbach have argued, a typically fascist move:

fascism is an extreme example of the political polarisation of gender with two basic types of bodies [which] exemplify the corporeal metaphysics at the heart of the fascist perception: the soft, fluid, and ultimately liquid female body which is [exemplified in the swamp and is] a quintessentially negative 'Other' lurking inside the male body ... [and] the hard, organised, phallic body devoid of all internal viscera which finds its apotheosis in the machine. (Benjamin and Rabinbach 1989, p. xix)

The essential characteristic of the machine is, as Michel de Certeau has remarked, 'male', though of course not biologically male (de Certeau 1986, p. 166). The existentialist marxist Sartre and the fascist soldier male seem to share a common gender politics.

In response to the project of capitalist modernisation in which 'all that is solid melts into air' and to the threat of communist revolution in which 'all that is solid becomes hot and liquid,' for the Fascist soldier male as Barbara Ehrenreich puts it: 'all that is rich and various must be smoothed over (to become like the blank facades of fascist architecture); all that is wet and luscious must be dammed up and contained; all that is 'exotic' (dark, Jewish) must be eliminated' (Ehrenreich 1987, p. xv). Or in the terms of the sublime, the beautiful and the slimy, all that is rich and various, wet and luscious, exotic and dark like the wetland must be given the smooth surface of the beautiful (like the canalised wet lands of Constable's wet landscapes) and not be rendered into the rough surface of the sublime, nor be permitted to remain the sticky, sucking surface, and depths, of the slimy and the swampy (see Figure 1).

The (al)chemical project of capitalist modernity takes the 'base matter', or 'slimy bads' as Zoë Sofoulis calls them, of the earth and transforms or sublimates it into 'commodities', or what Sofoulis calls 'shiny goods' of technology, then dumps its waste products back into the slimy, swampy places of the earth, and so the process comes full circle back, as it were, to where it began (see Figure 1). But in so doing the earth is destroyed to become like the landscape of trench warfare which Walter Benjamin saw as the terrain of German idealism and which can be seen as the final outcome of sublimation:

a fact to be proclaimed with all bitterness: it was only when the landscape was totally mobilised for war that the German feeling for nature could be so unexpectedly revived. This was a landscape originally settled by peaceful geniuses; but their sensuous vision was now eradicated. As far as the eye could see above the edges of the trenches, the land had become the terrain of German idealism itself – every shell-crater a philosophical problem; every barbed fence a representation of autonomy; every barb a definition; every explosion an axiom. By day, the heavens were contained in the cosmic interior of the steel helmet; by night, in the morality by which it was governed. Technology had attempted to retrace the heroic features of German idealism in burning fire and trenches. Technology was wrong: for what it took to be features of heroism were in fact the Hippocratic features of death. (Theweleit 1989, pp. 150–1)

The process of sublimation returns full circle, as it were, to the slime from which the whole process could be seen to have begun, so that it can begin again, though it has never stopped (see Figure 1).

IN THE SWAMP WITH EAGLETON

Critical attention to the gender politics of the sublime has been given recently by Terry Eagleton, though without discussing the gender and environmental politics of the metaphors he employs to do so whose burden is, as always, as remarked by Kay Schaffer, carried by women (Schaffer 1988, p. xii).[16] Metaphor is women's burden and Mother Earth's too. Eagleton shows how the sublime in the western tradition is associated with the masculine virtue of 'virile strenuousness'. Unlike the femininity of beauty, the sublime is 'on the side of enterprise, rivalry and individuation' (Eagleton 1990, p. 54). Yet the sublime and the beautiful do not co-exist side-by-side in blissful harmony. Rather, the sublime quite literally colonises the beautiful which, in turn, attempts to colonise the slimy. Eagleton argues that the sublime, 'the lawless masculine force . . . violates yet perpetually renews the feminine enclosure of beauty' (ibid. p. 54) which, in turns, encloses the uncanny, the slimy and the swampy (see Figure 1).

Yet enclosure is no mere metaphor. In the British enclosure movement of the eighteenth century, as Raymond Williams puts it, 'by nearly four thousand Acts, more than six million acres of land were appropriated' (though much of this land was specifically wetland, a crucial distinction) (Williams 1985, p. 96). Enclosure, argues E. P. Thompson, was 'a plain enough case of class robbery' (Thompson 1968, p. 237). The enclosure movement in wetlands was invariably associated with drainage in order to render the land profitable for monoculture with, as Christopher

Hill maintains, 'fen drainage alone extending the country's arable area by 10%' (Hill 1965, p. 260). In fact, the new outburst of drainage projects of the 1760s was contemporaneous with and supportive of the beginning of the agricultural revolution (Thirsk 1957, p. 205).

Enclosure is to the beautiful as drainage is to sublimation. Both enclosure and drainage were exercised against the slimy. For Freud the culture-work of his famous and oft-quoted pronouncement of 'where id was, there ego shall be' is, in the terms of his less quoted analogy, 'not unlike the draining of the Zuider Zee' (Freud 1964, p. 80; Theweleit 1987, p. 355). The draining of the Zuider Zee was probably the single largest drainage project ever undertaken, amounting to over half a million acres (Van Veen 1942, p. 52). The devastating effects of this project on the local people, their bioregion and livelihood are graphically depicted in the film *Doodwaater*, literally 'dead water'.

Dyking and draining the Zuider Zee produced dead water and no doubt the draining of the id produces the dead water of the dammed, and damned, wet land of the bourgeois ego unlike the living black water of the wild wetlands of indigenous cultures. Yet for Freud the id needs to be drained because it is 'a chaos, a cauldron full of seething excitations' which run into it from the instincts (Freud 1964, p. 73). If culture-work is ultimately draining instinct, then the threat of instinct for Freud is that it 'would break down every dam and wash away the laboriously erected work of civilisation' (Freud 1973, p. 353) and we would flounder about, as Lyotard puts it, in the swamp of uncertainty that fashions the instincts themselves (Lyotard 1993, p. 58 and 1989, p. 15).

For Kant, also, the sublime is associated with a bulwark against the conscious being swamped by the unconscious. The Kantian sublime is, Eagleton argues:

in effect a kind of unconscious process of infinite desire which like the Freudian unconscious continually risks swamping and overloading the pitiable ego with an excess of affects. The subject of the sublime is accordingly decentred, plunged into loss and pain, undergoes a crisis and fading of identity; yet without this unwelcome violence we would never be stirred out of ourselves, never prodded into enterprise and achievement. We would lapse back instead into the placid feminine enclosure of the imaginary, where desire is captivated and suspended. (Eagleton 1990, p. 90)

Or, what is worse, we would be immersed in the (s)mothering feminine slime of the unconscious, the feminine here split as a defence mechanism, as it was for the fascist soldier males, between the good woman in white and the bad, red woman, and even the black water of the swamp monster/mother.

These processes can be construed in historical and environmental terms, as Bachofen did to some extent, when the mountainous and phallic sublime of the capital(ist) city violates and colonises the beautiful feminine enclosures of English agriculture and threatens them with reversion to the slimy feminine of the swamps, the fens, the unconscious, the 'primitive' and the indigenous (see Figure 1). In response to this threat, the masculine sublime and the project of modernity maintain themselves by 'managing' in neo-imperialism the constant threat of inundation posed by the feminine and pre-modern slimy. Thus, Eagleton argues at length:

art for Hegel is not properly representational at all, but rather an intuitive presentation which expresses a vision rather than imitates an object. It incarnates a sensuous awareness of the Absoluto, abolishing all contingency and so showing forth *Geist* in all its organic necessity. As with Kant's aesthetic, this sensuousness is securely non-libidinal, freed from all desire, a beauty whose otherwise disruptive material forces is defused by the spirit which informs its every part. This intimate yet idealised body, material yet miraculously undivided, wordlessly immediate yet shaped and stylised, is what Romanticism terms the symbol, and psychoanalysis the body of the mother. It is therefore not surprising that Hegel finds something obscure and enigmatic about this form of materiality, something within it troublingly resistant to the translucent power of reason. Its betrays its unsettling force most obviously in the 'bad infinity' of oriental or Egyptian art, that grotesque spawning of matter, vague, groping and boundless, which threatens in its fantastic heterogeneity to swamp the pure spirit like some nightmarish creature of science fiction. Hegel finds such sublime proliferation of matter distinctly unnerving: within his system, this shapeless feminine stuff is redeemable only when impregnated by rational form, negated in its material presence and gathered into the inner unity of the Idea. The oriental stage of art is, so to speak, the child smothered by a fleshly mother; in the harmonious artefacts of ancient Greece and Rome, child and mother have achieved some symmetrical unity; at the highest, Romantic stage of art, when the almost disembodied spirit yearns for freedom from its material encasement, the child is in the process of definitively overcoming the Oedipal crisis and taking leave of the mother altogether. We do not, then, remain with the aesthetic long, but climb up a stage to religion, which still presents the Absolute in terms of images; and finally, if we can stay the course, we rise to the rarefied conceptual representations of philosophy itself. (ibid. pp. 143–4)

And so attain the sublime heights of capitalism, abstract theory and the superego in which patriarchal men can calculate themselves to be independent of nature. Repudiating their connections with their mothers, their own bodies, and with Mother Earth, they can continue to exploit the natural environment with impunity, a highly dubious, masculinist and self-destructive enterprise, though it represents the triumph of capitalist modernisation and of modernity – not least over wetlands.

NOTES

1. Elsewhere Lyotard (1989, p. 200) has argued that 'the sublime is perhaps the only mode of artistic sensibility to chatacterise the modern'.
2. For the abject see Kristeva 1982.
3. I am grateful to Alec McHoul for drawing my attention to this earlier work of Kant's. I am also grateful to Stephan Millett for his comments on my discussion of Kant's work.
4. I am grateful to Ann McGuire for pointing out the pertinence of Spengler's work to mine.
5. I am grateful to Zoë Sofoulis for allowing me access to her thesis and to Maria Angel for pointing out its pertinence. I am also grateful to Zoë and to Susan Hayes for making comments on an earlier version of this chapter.
6. Exceptions include Daly 1979, pp. 332, 390n and 402; and Bauman 1973, pp. 137–42. I am grateful to Jon Stratton for drawing my attention to Bauman's discussion. A more recent exception is Collins and Pierce 1976, pp. 112–27. I am grateful to Ron Blaber for drawing my attention to this paper. The most recent exception is Harvey 1991, pp. 49–66. I am grateful to Anne Brewster for drawing my attention to this article. To complete the genealogy, an earlier version of the present chapter appeared in *New Formations*, 18 (1992), pp. 142–59.
7. Theweleit has described Bachelor Births from Bachelor Machines as 'attempts to create a new reality by circumventing the female body – to engender the world from the brain' (Theweleit 1987, p. 330) and from the head as the site of the superego (ibid. p. 414) both exemplified in Zeus giving birth to Pallas Athene from his head, arguably the first Bachelor Birth from a Bachelor Machine and the first 'brain-child' (see Figure 1). See also de Certeau 1984, pp. 150–3 and 1986, pp. 156–67; and Carrouges 1954.

8. See Hebdige 1987, pp. 47–76. Against the sublimes of Kant and what he calls the 'post avant-gardes' Hebdige valorises the vulgar, the popular and 'the body' (though whose body? one might ask). Yet repugnance for or valorisation of the vulgar, the popular and 'the body' can be regarded as setting up a sterile antinomy which occludes how the sublime is supported precisely by sublimation of the vulgar, the popular and 'the body'. Part of the impossibility of the sublime object lies not only, as Hebdige suggests, in being in a sense beyond the fetishisation of commodities, but also in its production from raw materials by labour and its imbrication in the circuits of sublimation. The vulgar, the popular and 'the body' do not have a fixed and necessary meaning or resistive force against capitalism (early or late), but are sublimated in the production and consumption of commodities.

9. For the sight of a mountain as evoking a feeling of the sublime see Kant 1960, p. 47 and 1952, p. 121. Similarly for Thoreau (1980, p. 44) 'sublimity and grandeur . . . belong to mountain scenery'. For 'mountain sublimities' etc. see also Muir (1911) 1987, pp. 110, 111, 115, 122, 140, 141. The quintessential pictorial evocation of mountainous sublimity is 'The Wanderer' by Caspar David Friedrich with its individualist nomad/monad astride the pinnacle surveying the cloud swept, sea-like heights. This painting is reproduced on the cover of Eagleton 1990.

10. Quoted in 'slime', *Oxford English Dictionary*, p. 706.

11. I am grateful to Bethany Brown for pointing out to me the pertinence of Paglia's book.

12. I am grateful to Rita Felski for alerting me to the fact that Bachofen's work was available in an English translation and that a copy was housed in the Murdoch University Library.

13. I elaborate the discussion of the wetland as placenta (and vice versa) in Chapter 6.

14. This list is based on and adds to those of Cixous and Clément (1986, p. 63) and Poynton (1985, p. 18).

15. I am grateful to Susan Hayes for pointing this out to me.

16. Irigaray likewise refers to the 'masculine' games of tropes and tropisms' in which women are 'rolled up in metaphors, buried beneath carefully stylized figures', though 'no metaphor completes her' despite the fact that 'the womb has been played with, made metaphor and mockery of by men' (Irigaray 1985, pp. 140, 144, 229 and 263). See also Derrida 1982, pp. 207–71.

II

Cities and Swamps

3

A Modern City and its Swamp Sett(l)ing: Decolonising Perth's Wetlands

The relationship between cities and wetlands has been a close but vexed one, especially in modern times: some cities, such as St Petersburg, were built on land reclaimed from marshes, whilst some new colonial settlements which grew into cities, such as Perth (Western Australia), were founded between a river and swamps. These cities could only expand uninterruptedly by filling or draining the swamps. The slums of other cities, especially industrial ones, were built on marshes: Moss Side in Manchester, the Bogside in Londonderry (even the names are indicative of their wetland sites), and much of the East End of London (Purseglove 1989, p. 27). In this chapter I trace the early colonial history of Perth and its colonisation of its wetlands. My title alludes to and elaborates on that of *A City and its Setting* by George Seddon and David Ravine (Seddon and Ravine 1986). Yet whereas they see a city situated benignly in its setting, I see a city malignantly settling its swamp setting, settling down to make the setting comfortable, to render the uncanny and unhomely place homely. In the process wetlands were destroyed or degraded.

The project of colonisation, especially in its modern phase and especially in relation to the establishment of settlements and the foundation of cities, is strongly tied to the drainage or filling of wetlands; in fact, the latter makes possible the former. Modern cities sublimate the slimy or swampy (see Chapter 2, Figure 1). Their skyscrapers raise the city above the swampy; their height represses the depths; their suburbs spread out, filling the wetlands, or more recently, aestheticising them in accordance with the conventions of English parkland. As far as some modern cities go, the project of modernisation has been to fill or drain wetlands. Wetlands have posed and still pose an obstacle to urban development. In order for development to take place wetlands are filled in order to create a smooth and replete surface on which development can then take place.

FOUNDING A SETTLEMENT IN A SWAMP

Like its perhaps more illustrious counterpart in St Petersburg located on the marshy delta of the Neva (literally 'Mud') River, the layout of Perth on the banks of the Swan River was constrained by geographical conditions. Also, like St Petersburg with its fortresses, its location was partly the product of military considerations (Berman 1983, pp. 176–81 and Dluhosch 1969, pp. xiii–xxix). The close association between war and the city has been remarked on by Paul Virilio for whom 'the city is the result of war, at least of preparation for war' (Virilio and Lotringer 1983, p. 3), not least the war against wetlands. Cities were built first for defence against aggressors or competitors, but in order to establish themselves they werre aggressive towards the natural world. Once established, the city is the head-quarters for the war against the base waters of wetlands. Perth is no exception to this general rule including the initial choices made for its site. Historians have wondered why Captain James Stirling founded the settlement of Perth, the capital of the new colony of Western Australia, seven miles up the Swan River from the port of Fremantle at its the mouth. One explanation advanced by D. C. Markey is that Stirling chose the site for the purely military reason that it could not be bombarded from the sea and that the river was not navigable by gun sloops (Markey 1979, p. 347).

Yet the site which Stirling chose also had another advantage as far as military considerations were concerned. Located on a narrow ridge of land between the Swan River to the south and an interlocking network of swamps and lakes to the north, it was protected from land attack from both directions. Markey argues that the situation was 'strategically sound for, though within reach of the port, it was inaccessible to a waterborne hostile naval force, and a land attack from the north would have floundered to a halt in the morass of lakes and swamps' (ibid. p 351). It is possible to see the city as founded on a dialectic of attack and defence, aggression and paranoia, a sublime vigour and a cowardly quailing, a forward and phallic thrusting into new territory and a defensive desire to protect its own rear through its swampy backblocks. Relatedly, the city is also founded on a dialectic of accessibility and isolation: neither too vulnerable to be open to attack, nor too isolated to be inaccessible, but on the edges of the empire, colonising the unknown.

Yet what may have been strategically sound for military reasons was not necessarily, and in this case certainly not, strategically sound in terms of town planning and urban development. The lakes and swamps to the north were a hindrance not only to military invasion but also to urban expansion and, according to Markey, the topography dictated an initial, unusual linear development (ibid pp. 346 and 351). As a result the main thoroughfares ran east–west. However, as no serious military threat was ever posed to the capital of the new colony the hindrance to urban expansion could be quickly overcome.

Whilst military considerations may have been uppermost in Stirling's mind in choosing a site, aesthetic considerations were, nevertheless, taken into account. Two weeks after the foundation ceremony in 1829 Stirling wrote to the Parliamentary Under-Secretary for the Colonies that 'the position chosen . . . is one of great beauty as respects scenery' (ibid. p. 348). Presumably, though, the position of the settlement with swamps to the north was not of great beauty as respects scenery. Nor was it probably for a visitor to Perth in 1829 who wrote that Perth has 'one of the most delightful demipanoramic views, I suppose, in the world' (Seddon and Ravine 1986, p. 62). No doubt this view was to the south across the river rather than the north across the swamps.

The view across the swamps was certainly not beautiful nor delightful for George Webb in 1847 who compared Lake Monger, one of the lakes north of Perth, to the lakes 'at home' and found it not only wanting but also repulsively unhomely (*unheimlich*, or uncanny, in Freudian terms):

at home, a lake is known only as a sheet of water which seldom or ever is dried up, and it is naturally associated in one's mind with pleasant and picturesque scenery, but here it is quite different . . . there is an air of desolation about these lakes which strikes the spectator at once . . . It is complete still life without one point of interest in it, as far as striking scenery goes, and totally different from anything I ever saw outside Australia. (Webb 1847, pp. 161–2)

Perth's lakes are here marked negatively as 'quite different', 'totally different' and as lacking 'interest'. Indeed, Webb himself notes 'the very marked difference between those I had ever seen at home and the Lagoon or Lake I am now speaking of (ibid. p. 161). Incapable of dealing with scenic and geographic difference on its own terms, he constantly reverts to what is old, familiar and long known. This difference is construed in relation to water as a datum point – whether the lakes dry up or not, whether they are a sheet of water or not – and in relation to scenery with its mobilisation of such aesthetic categories as 'pleasant', 'picturesque' and 'striking' which are applied to the lakes 'at home'. By contrast, Perth lakes are 'a desolate still life with no point of interest'.

The new, unfamiliar and unknown is construed as alien, to be denigrated and despised in normatising absolutes which set up a clearly preferred position for reading. Other, alternative readings of the (wet)landscape are effectively excluded and relegated beyond the pale of sophisticated aesthetic judgement. A similar kind of judgement to Webb's was made some years later by George Moore who recorded in his Diary that 'I cannot compare these swamps to any marshes with which you are familiar; perhaps a tract of ground covered with old willows and green weeds, with here and there open spaces of deepest water, is the nearest resemblance I can supply' (Moore (1884) 1987, p. 157).[2]

Interestingly, and symptomatically, there seem to be few surviving visual representations of the swamps and lakes to the north of early Perth. A painting such as Horace Samson's 'Perth in 1847' (Figure 1), in which the swamps and lakes may have figured in the background (though many would have been drained by this time), merely recedes into English-style parkland. A drawing such as Charles Wittenoom's 'Sketch of the Town of Perth from Perth Water' (Figure 2) published in 1839 as the frontispiece to Nathaniel Ogle's The Colony of Western Australia: A Manual for Emigrants, does show a large lake (probably Kingsford Lake) in the left midground suitably aestheticised for the contemplation of new or potential settlers.

The verbal representations of Webb and Moore contrast strongly with the visual representations of Samson and Wittenoom. All employ, and all are constrained by, the rhetoric of their media, by the conventions of depiction of the wetlandscape. The verbal means had developed an elaborate, albeit denigrating, vocabulary, whereas the visual was locked into the aestheticising conventions of English landscape painting. Whilst American painters during the nineteenth century were developing a rhetoric of visual representation in order to depict the complexity and lushness of American swamps, there seems to have been no comparable development in nineteenth-century Australian painting.[3]

Although the lakes and swamps around Perth were initially perceived as aesthetically displeasing, they were also, by contrast, perceived as having an economic potential and usefulness, at least by some early writers and settlers, though not by others. In the 'Introduction' to the collected journals of several expeditions made in Western Australia during the first four years of the colony, 'lagoons and salt water lakes' are described as being 'scattered everywhere' and although the former do not seem to worthy of any mention as possessing economic value, 'the latter may be looked upon as depositories from whence wealth and commercial eminence may hereafter be derived to the settlement' (Journals (1833) 1980, p. xiv). What these economic benefits might be the writer does not specify, and why the lagoons seem to have none he does not say.

Yet the lagoons were exploited economically in the early days of the colony by Samuel Kingsford who 'proposed the ambitious scheme of draining the swamps and lakes to the north of Perth' in order to provide the motive power for his mill (Hasluck and Bray 1930, p. 80). Under his proposal he wanted, in the words of a contemporary, to

cut a deep trench and lay a pipe from some lagoons behind Perth into the town to afford him a supply of water. There are some of these lagoons eight miles in circumference, and at no great distance, which he thinks have a communication with each other through the sandy soil, or which may be made to communicate with inexpensive cuts. (ibid.)

Certainly these lagoons did communicate with each other via the underground aquifer though some early maps actually show them joined together on the surface (Figure 6). But presumably Kingsford was only concerned with these hydro-geological niceties in order, as Raymond Williams puts it, 'to make Nature move to an arranged design' (Williams 1973, p. 124).

The colonial Government was certainly sympathetic and more than generous when Kingsford approached it with his proposals. He was given

the perpetual right of converting to the use of the mill the water of Kingsford Lake, Irwin Lake, Lake Sutherland, and Lake Henderson and it was agreed that the use of the water of Monger's Lake and the Great Lake (that is, Herdsman's Lake) should not be granted to any one else for five years, by which time Kingsford would have had an opportunity of testing whether he wanted more water for his mill . . . Kingsford was also given rights to use as much land as necessary to connect up these lakes, making of them a common reservoir. (Hasluck and Bray 1930, p. 80)

This was a very generous grant indeed, and not only for the present. Effectively the waters of Monger's and Herdsman's Lake were being given as free 'futures'.

The lakes and swamps to the north were not only an impediment to development and a source of water power. They also posed a threat to the health of the population in that they 'bred huge numbers of mosquitoes and other pests' (Markey 1979, p. 351). In 1847 the Colonial Surgeon, Dr J. Ferguson, wrote a letter to the Colonial Secretary:

I have the honour to acknowledge receipt of your communication of yesterday's date acquainting one 'that it had been represented to the Governor that the accumulation of water in the Swamps in the back streets of the Townsite of Perth, has a tendency to generate Fever and other diseases in that neighbourhood and to aggravate the Violence and obstinacy of such diseases' and conveying His Excellency's desire 'that I should investigate the subject and report whether I have any reason to suppose that the existence of these swamps is prejudicial, or likely to be so, to the health of the Inhabitants of the Town in general, or to the residents of that part of the Town in particular and if so whether I consider the draining of the water from the swamps – / and whether wholly or partially / – would be likely to be attended with any beneficial results to the public health.'

I have to express my regret that my time / having to give in my report this day / did not admit of my investigating this subject to any great extent, and that my short residence in Perth does not enable me to say *much* from personal experience – but I unhesitatingly give it as my opinion that the existence of these swamps is prejudicial to health, and much more likely to become more so as the population of the Town increases – and, as a matter of course, I consider that the complete draining – / I do not consider that the partial draining would remedy the evil in any degree / – or filling up of these swamps would be attended with any beneficial results to the public health.

I beg, however, to state that in giving this as my opinion, I do not pretend to say how far the draining of the water from the swamps may effect the wells in the Town – a scarcity of good wholesome water during the hot summer months would be a serious evil, and as likely to be attended with grievous consequences to the health of the inhabitants of Perth as any danger at present resulting from the existence of these swamps.[4]

Ferguson tries to balance what he sees as the lesser evil of the threat to public health posed by the swamps themselves against the greater evil presented by a potentially unhealthy water supply as a possible consequence of draining or filling.

The problem of the threat the swamps posed to health continued to persist as the 'Inspector of Nuisances' reported in 1869 that Perth in the last few years 'had witnessed a great deal of sickness, with many cases of local fever occurring, some of them fatal.' Yet he was pleased to report that in the past six months there had been less sickness in the city, perhaps due to his own efforts. But the drainage of the city was very defective: 'so long as bad drainage exists we are liable to periodical attacks, more or less severe, of sickness' (Stannage 1979, p. 162). Considerably more, if we were to believe the Reverend Meadowcroft who in 1873 warned that 'if effectual drainage was not adopted quickly' then 'the city would probably be visited by some dire epidemic' (ibid. p. 169). Such warnings were the stock in trade of the nineteenth-century Sanitary Movement which I discuss in greater detail in Chapter 5.

Yet it was perhaps more the demand for space for market gardens, rather than the desire to improve the health of the population, which gave the final impetus for drainage. This demand for market garden space was partly the product of population pressure as C. T. Stannage has argued:

with a population which increased nearly five-fold between 1850 and 1884 Perth needed a vastly increased supply of food and drink. This accounts for the big increase in the number of market gardeners in the town . . . This growth was possible only with the draining of the swamps behind the town. While there were some winters when the drainage system failed and the gardens were inundated with water, and while the drains had to be cleared of debris regularly, by the 1870's Perth was surrounded by gardens in a fan ... (ibid. p. 128)

Over sixty years after the drainage of the wetlands behind the town the then Town Clerk of Perth was moved to regret the loss of what he called:

the opportunities which presented themselves at that day of laying out an ideal garden city by taking advantage, for ornamental purposes, of the chain of lakes ... The early settlers, however, were probably influenced more by the possibilities of the site from an agricultural point of view. (Bold 1939, p. 30)

And certainly not from an aesthetic point of view. In response to Bold's (bold) vision Geoffrey Bolton has remarked that 'such planning would have required financial and engineering resources far beyond the reach of an impoverished colony struggling for survival' (Bolton 1989, p. 144). But if this opportunity had been taken, as George Seddon has postulated, 'Perth would then have been surrounded by open space on all four sides, like Adelaide's Parklands' (Seddon 1972, p. 230) which were, however, designed and laid out on the English model.

Instead of Perth becoming an ideal garden city surrounded by public open space, it became an ideal market garden city ringed by land privately owned and developed for the production of vegetables. In 1980 Su-Jane Hunt claimed, nevertheless, that it was a Garden City (Hunt 1980, p. v), though certainly not in accordance with Bold's vision. Here the land (including wetland) was being organised, as in England, and as Raymond Williams remarked, for production and consumption (Williams 1973, p. 124). Indeed, as Alexandra Hasluck puts it, though perhaps not with the same degree of opprobrium which now attaches to the terms, Perth and the Swan River settlement were 'a colony for capitalists (Hasluck 1956, title of chap. 2).

What would Perth look like today if the lakes and the swamps to its immediate north had not been drained? What if Colonel Light, who planned Adelaide's open Parklands (though governed by military, rather than recreational or aesthetic, reasons[5]) had overseen Perth's development instead of John Septimus Roe,

Western Australia's first Surveyor-General (Jackson 1982, pp. 58–60)?[6] Would there today be a network of lakes and swamps set in acres of parklands and crossed at the narrowest points by raised roadways to allow vehicular and pedestrian traffic through them for business communication and for leisure?

The wetlands to Perth's immediate north have largely disappeared. But they have, nevertheless, left a trace in that indigenes living in the area still move around in it using these lost wetlands to orientate themselves.[7] Beneath the level of the streets and buildings spread out over the surface of the now lost wetlandscape, filling up its hollow places, smoothing out its wrinkles, draining its damp places, another, different positioning still goes on in which these people orientate themselves to their lost wetlands. This orientation and movement could be nostalgia for a bygone pre-invasion era, or an attempt to recover a mythic lost origin in relation to the (wet)land, or an assertion of a resistive reading of the townscape which reads it in terms of the lost wetlandscape beneath it or simply locating and mourning the loss of a destroyed place of sustenance, or all four.

The wetlands of the Swan Coastal Plain continued to be assessed in terms of a capitalist imperative into the twentieth century. Yet whereas they were seen to be profitable for market garden use in the nineteenth century (and even after being reclaimed, are still seen as such because of their proximity to groundwater supplies), to a writer in 1905 they had no economic value whatsoever: 'between the Darling Range and the coast there are a few salt-water lagoons, and many small fresh-water lakes, the majority of which are nothing more than swamps during the dry season, and none of them are of any economic importance' (Bekle 1981, p. 21). Precisely as for George Webb and Lake Monger, these wetlands are strongly marked in pejorative terms: 'nothing more', 'none of them'. Yet whereas Webb construed that negativity in aesthetic terms, this writer does so in economic terms. The categories of swamp and economic importance are deemed mutually exclusive. Swamps, then, could be filled with impunity.

Seven years later J. S. Battye generalised in very similar terms of the entire State:

there are no lakes worthy of the name throughout the State. As regards the
great interior, this may be explained by the arid nature of the country due to
the slight and irregular rainfall, and the rapid evaporation of moisture caused
by the intense heat of the sun. The so-called lakes of this great region, in many
cases considerable in extent, are, except after occasional heavy rains, merely
immense salt marshes or claypans. But even between the ranges and the sea,
where we might expect to find sheets of permanent water, there are very few,
and these, except during the rainy season, are little better than swamps or
marshes. Salt lagoons occur in place, but have little or no economic value.
(Battye 1912, vol. I, p. 2)

Battye perpetuates the colonist's mentality. Going back through
Webb to the early founders and explorers, he compares Western
Australian wetlands unfavourably to the lakes 'at home' and
denigrates their aesthetic and ecological difference. Lakes are
implicitly privileged over claypans and marshes, but if the latter
are called the former then they are for Battye 'so-called lakes' and
'merely' marshes or claypans.

Against the litany of the pejorative it is possible to counter with
some affirmative press for Perth wetlands. Yet this seems to be the
exception to the rule. In one instance positive press is marked by
surprise, if not incredulity. In a newspaper report in 1911 a
Christmas Tree fête held at Forrestdale Lake is evoked in the
following terms of a pastoral idyll: 'though the day was warm, the
breeze blew cool across the lake and the visitors were delighted
and surprised to find such a beautiful picnic ground, and spent
the hour before tea by the water' (Popham 1980, p. 107). The
cooling effect of evaporation from wetlands is one of the few
positive qualities to have traditionally been associated with them
in western culture. One lake in the wheatbelt region of Western
Australia is remembered as having a dense canopy of sheoaks and
paperbarks right through the lake. Walking underneath it on a hot
summer's day was 'like walking into a refrigerator' (Sanders 1991,
p. 10).

MAPPING THE CITY AND THE SWAMPS

Unlike many cities whose beginnings are shrouded in 'the mists of
time', Perth is not only a city whose beginnings can be traced but

also a city which was founded where no previous city or settlement had existed. Perth is, as George Seddon and David Ravine
argue,

typical of the Australian capital cites in being a 'plantation' city, externally
imposed on a new environment, but such cities are unusual in their origins
compared with most of the world's cities. Melbourne and Brisbane [however]
began their growth, at least in part, from indigenous economic and geographic
forces, but the others were established at one blow by policies and money
derived from the other side of the world. The decisions were taken on grossly
inadequate information derived from cursory local surveys. (Seddon and
Ravine 1986, p. 62)

As a plantation city and as a city founded on a narrow strip of
land between a river and swamps, Perth was not in many respects
a typical colonial city, though it was in others. For Franz Fanon,
the pioneer theorist of decolonisation, the archetypal colonial city
was divided between the Settler's Town basking on the hillsides
and the Native Town 'wallowing in the mire' as it does, for
example, on the mud flats of the Ganges in Chandrapore (Fanon
1967, p. 30; Forster 1936, p. 9). Although the backside of Perth
wallowed in the mire of its swamps, there was no Native Town
there and the hillside of Mount Eliza was too steep for the settler's
town or the 'civil station'.

Plantation cities like Perth took place, as far as Europeans were
concerned, on a *tabula rasa* in a *terra nullius*, on a clear sheet in
an empty land, though in order for the northward expansion of
Perth to take place the *tabula* was a sheet of water, especially
during the floods of 1842 and 1847 (Stannage 1979, p. 60). Perth
was also unlike many colonial cities in that it was, as D. C. Markey
indicates, built to a plan and in this regard was remarkably
different from most of the other towns developing at that time in
the other Australian colonies (Markey 1979, p. 351), except perhaps Melbourne and Adelaide.

There is a certain cultural contiguity between the aquatic or
terrestrial, or more precisely aqua-terrestrial in the case of Perth,
tabula rasa of the colonised (wet)landscape and the manuscriptic
tabula rasa of the colonial plan. The map itself is a powerful
instrument of colonisation. Indeed, Michel de Certeau goes so far
to argue 'the map colonises space' (de Certeau 1984, p. 121). Or

put more simply and bluntly, 'maps are territories' (Turnbull 1989). Maps are territories and maps colonise space because, in Paul Carter's words, the map is 'an instrument for performing geometrical divisions . . . [which] reconciles all viewpoints into one unifying cause-and-effect perspective' (Carter 1987, p. 204). The map records in the scalar grid of longtitude and latitude. It accords the map reader a God-like, transcendent position above everything, seeing everything simultaneously unlike perspective which centres everything on an eye/I. In cartography everything on the surface is laid out for the gaze, unlike perspective 'which already mummifies life,' to use Luce Irigaray's words (Irigaray 1991, p. 115). In the map every place becomes space, and every time, eternal. This means that for wetlands their distinctive ecosystems are reduced to outlines on a map, if indeed they are mapped at all.

Mapping of wetlands poses a problem because of their seasonal fluctuations in depth, and hence in surface area. A map indicates surface and not depth; it reduces depth to surface, and only at a certain moment in time. The map is synchronic not only because it is drawn at a particular moment in time, but also because it freezes everything in that moment of time. Wetlands, however, are highly diachronic and spatially variable unlike a range of other landscape features, such as ranges. Wetlands require recognition of their diachronicity. A recent attempt to map Perth wetlands sponsored by the Water Authority of Western Australia begins by classifying and categorising wetlands in terms of their temporality, or water permanence (whether they are permanently or seasonally inundated or waterlogged), and in terms of their spatial configuration, or cross-sectional shape (whether they are basins, flat, channels or estuarine) (Figure 3). This map also establishes the current stock of wetlands in the region against which we can compare their earlier absence from maps.

The early maps of the Swan River region, such as those by de Vlamingh of 1697 (Robert 1972), by Freycinet and Heirisson of 1801 (Appleyard and Manford 1979; Marchant 1982), and by Stirling himself of 1827 (Figure 4), do not show the lakes and swamps of the region at all. They are completely absent. It is this very absence which constitutes their initial colonisation. They are *aqua-terra nullius*. Not even seen or acknowledged in the official discourse of the map, they are just like the 'native' inhabitants who were regarded by Cook as having nothing to do with the state of the country as he found it (Wright 1990, pp. 3–7). Hence, there is not only a certain cultural contiguity between the map and the wetlandscape, but also between the absence of wetlands from the colonial map and the absence of reference in Cook's journal to the work of indigenes in transforming the environment. Both are beneath the level of visibility, out of sight and out of mind, repressed and oppressed.

The map is not as totalising (though it is still as colonising) as it might seem since it only records either what is seen (perhaps de Vlamingh, Freycinet, Heirisson and Stirling did not see the lakes and swamps either because they were so fixated on the river or because they did not conform to their idea of what lakes should be) or what the map-maker wants to be seen in order to promote further colonisation and settlement (and so a map of 1829 of the very new settlement of the Swan River does not show the lakes and swamps either on the entire Swan Coastal Plain (Figure 5)). Of course, it would have involved a massive surveying operation to have plotted even the larger lakes and swamps of the region, though this effort has been taken, albeit to no great extent, in maps of 1833 of Perth and the region drawn at different scales (Figure 6). The different scales draw certain size objects into their purview and exclude others. The map of Perth shows a large, interlocking network of lakes and swamps to the north of the settlement and the map of the region shows what must be Forrestdale Lake as a 'Large Lake' covering an area about twice its present size. Even the fact that this Lake was mapped at all was probably only the result of the fact that explorers happened to stumble on it, and were obstructed by it, in their quest to find the source of the Canning River.[9]

Although the wetlands of the Swan Coastal Plain were not, perhaps understandably, marked on maps, it would have taken little effort to have described them. Stirling does not make this effort on his map, though he describes the Swan Coastal Plain as 'undulating grassy country thinly wooded' (Figure 4). Wetland-scape is reduced to landscape, albeit English pastoral landscape. It would not have promoted colonisation and settlement if Stirling had written 'swampy country thickly wooded'. The wetland areas to the north and south of the Swan River are a blank space and an empty land, a *tabula rasa* and a *terra nullius*, as they are on the map of the new settlement. This view was part of what Suzanne Falkiner has recently called 'the geographical and philosophical blank slate that Australia presented to the European mind' (Falkiner 1992, p. 10).[10] On all four maps the wetlands of the region are conspicuous by their absence. This absence of wetlands from maps precedes, makes possible and justifies before the event their actual removal from the land, much as Cook's absence of all reference to the work of indigenes in transforming the environment legitimated genocide. Perhaps we need a term comparable and analogous to 'genocide' to apply to the draining and filling of wetlands, such as 'aquaterracide,' the killing of wetlands.

The colonial map is distinguished by the politics of presence and absence, by what gets plotted or named and what does not. Either way, the colonial map reduces space, and the earth, including wetlands, to a flat surface of inscription. In and by this process the depths of the wetland are repressed. Surfaces of inscription, Lyotard argues, are 'themselves flows of stabilised quiescent libidinal energy, functioning as locks, canals, regulators of desire, as its figure-producing figures' (Lyotard 1984, p. 98). The reduction of the depths of the earth in general and of wetlands in particular to the surface of the map and the surface of the earth represses the nether(wet)lands of the unconscious. Yet this reduction of depth to surface is, Gilles Deleuze and Felix Guattari argue, a necessary condition for social reproduction: 'some kind of full body, that of the earth or the despot, a recording surface, an apparent objective movement, a fetishistic, perverted, bewitched world are characteristic of all types of society as a constant of social reproduction' (Deleuze and Guattari 1977, p. 11). In the case of wetlands, the earth is the dry land, constituted as full body by not having its wetlands marked on maps and/or by having its wetlands drained or filled.

Wetlands are the 'unfull' or lacking body of the earth. As a consequence of this process of producing a full body, the earth is, Deleuze and Guattari go on to argue, 'the surface on which the whole process of production is inscribed' (ibid. p. 141). Although the earth may be covered with water, it is still a surface for the inscription, or writing, of colonisation and development. Writing for de Certeau is:

the concrete activity that consists in constructing on its own, blank space (*un espace propre*) – the page – a text that has power over the exteriority from which it has first been excluded . . . it allows one to act on the environment and transform it . . . it is capitalist and conquering . . . And so is the modern city: it is a circumscribed space in which . . . the will to make the countryside conform to urban models is realised. (de Certeau 1984, pp. 134–5)

The page of the colonial map has power over the wetlands which it has excluded. It constitutes the wetlands as object, or more precisely abject (an aspect of wetlands to which I return in later chapters) to be managed and transformed.

Although writing has power, this power is relative since writing is subject, in turn, to the power of speech. In a number of places Jacques Derrida has argued along the lines that 'the history of truth,' or at least the western history of (western) truth (there is a high degree of lexical redundancy here), 'of the truth of truth, has always been . . . the debasement of writing, and its repression outside "full" speech' (Derrida 1976, p. 3). As a result, writing has become synonymous, or at least associated, with other 'western' repressions. Writing, in Derrida's terms, 'the letter, the sensible inscription, has always been considered by Western tradition as the body and matter external to the spirit, to breath, to speech, and to the logos' (ibid. p. 35). In terms of colonialism, writing is the messy and corporeal colonial process, the 'white man's burden', external to and repressed by the civilisation and enlightenment of the colonising nation yet necessary for and to it. In terms of the natural environment, the wetland, the swamp, the marsh, has invariably been considered by the patriarchal western tradition as the lower body and dead matter, the grotesque lower bodily and earthly stratum, external to writing, to the mark, the inscription and doubly external to speech, spirit, and to the colonial centre and seat of power.

Yet the colonial process of writing does not master a compliant and passive object but tries to master an active agent which is a threat to the colonising power, particularly in the form of the wetland. The repression of writing is referred to elsewhere by Derrida 'as the repression of that which threatens presence and the mastering of absence' (Derrida 1978, p. 197). The colonies threaten the presence of the 'home' power and its mastering of the absence of home and the homely in the colonies, especially in the unhomely swamps. This threat is posed by the return of the repressed. The empire writes back, the swamp strikes back, by leaving traces on the surface of the clean and proper body of the surveyor (William Byrd) or the philosopher (Sartre) by immersion in the slimy and swampy. Writing as inscription on the surfaces of the body and the (wet)land is the instrument of colonisation; writing as trace on the surface and in the depths of the body, the slimy and the swampy is a threat to colonisation. There are not two sorts of writing, but writing is double, split between the arche-writing of the trace and the colonial writing of the inscription. Both always co-exist and take place on the surface of the body and the land. The writing of the map masters the absence of home and the map, in turn, is mastered by the speech of the colonial power. But the unhomely repressed threatens the homely and always returns in cultural dreaming and parapraxes.

A distinction needs to be drawn between what Derrida describes as 'the mastering of absence as speech and the mastering of absence as writing' (ibid.). Yet the two masteries are linked and operate in conjunction with each other. Mastery of speech masters writing; speech's mastering of absence masters writing's mastering of absence. Explorers and mapmakers master absence, the absence of 'home', of Europe, by writing journals and drawing maps which master the colonies. This writing is mastered, in turn, by the speech of colonial administrators and secretaries who doubly master the absence of home in the colonies mediated through the colonising process of map and journal. The colonial power holds the colonised lands at arm's length by articulating and mediating its power through writing. Besides a double process of mastery, there is also a double process of repression in which speech represses writing which in turn represses the land in general and the wetland in particular. As writing is the repressed of speech, so the swampy and the slimy are the repressed of writing. They threaten to return and to mark the body of the writer. Perhaps the slime and the swamp are such powerful figures for the unconscious because they are doubly repressed (see Chapter 2, Figure 1).

The modern colonial city is also implicated in and complicit with these articulated, double masteries and repressions. The modern colonial city is made possible by the colonising writing of the map and it is itself a colonising writing on the surface of the earth. It transforms wetlands into market garden and market garden into suburb thus completing the transition from wetland through agri-culture to suburbi-culture with the possibility of aqui-culture aborted. The colonial map and modern city, colonialism and modernity, are all writing. Modernity is colonialism at home; colonialism is modernity away from home. Modernity colonises and colonialism modernises lands, especially wetlands. For François Furet 'modernisation, modernity itself, is writing' (de Certeau 1984, p. 168). The map, the modern city, modernity and colonialism all write on surfaces, the surfaces of the earth, the surfaces of bodies, the surfaces of the wetland. In Graham Swift's novel *Waterland* Crick the narrator asks of prosperous maltsters whether they see 'in these level Fens – this nothing landscape – an Idea, a drawing-board for your plans' (Swift 1984, p. 15). Something similar could have been asked of the prosperous market gardeners of pioneer Perth.

Maps of early colonial Perth in 1833 (Figure 6) and in 1838 (Figure 7) trace the cartographic colonisation of its wetlands which went hand-in-hand with their actual settlement. These maps shows how what Carter calls 'the rational principle of the grid' was used to produce what he also calls 'the grid-plan town' which was, like the map which preceded it and made it possible, 'paradoxically placeless and directionless' (Carter 1987, pp. 205). The grid-plan town was derived from the military camp which produces what Henri Lefebvre calls 'an instrumental space . . ., a rectangular, strictly symmetrical space . . . with its strict grid' (Lefebvre 1991, pp. 244 and 245). It also produced what he goes on to call 'the cult of rectitude, in the sense of right angles and straight lines'. Through this cult 'the order of power, the order of the male is thus naturalised' (ibid. pp. 305 and 361), and the female order of the swamps alienised.

The grid achieved this power because it is for Carter 'the supremely historical figure', or perhaps more precisely the supreme historical figure. On the one hand, the grid was 'a means of speeding up the appearance of things, for hastening the nearness of distant objects', and on the other it 'shared the qualities of the explorer's track and the appeal of the picturesque view'. Both are necessary for capitalist 'progress'. 'The rational grid plan' was not only associated with 'the myth of progress' but was what Carter calls 'the matrix of physical progress' (Carter 1987, pp. 210, 215, 219 and 221) itself, capitalist 'progress'. The Swan River colony was a colony for capitalists because, as Carter puts it, 'the rational and equal division of land into purchasable blocks was the essential precondition of capitalist settlement . . . what the grid generated was wealth' (ibid. pp. 203 and 212). The grid was a machine for producing private property and so private wealth.

The grid had a profound and devastating effect on Great Mother Earth, the *magna mater*, especially on her most fertile and vulnerable places, such as swamps. As marsh and modern ways are inimical to each other, so are matrix and *mater* (despite the fact the former is derived from the latter though as a mathematised abstraction). Indeed, as matrix and modern ways go hand in hand, so do marsh and *mater*. Modern ways and matrix are inimical to marsh and *mater*. The grid is also the supreme historical figure of the divide between nature and culture because, as Keith Thomas puts it, 'neatness, symmetry and formal patterns had always been the distinctively human way of indicating the separation between culture and nature', or more precisely the patriarchal way of indicating the separation, and not the mere separation as if both could then co-exist in harmony, but the instrument of colonisation of nature by culture (Thomas 1984, p. 256).

The gird is composed of straight lines whose power is not just confined to the rational principle of the grid nor to the grid-plan town with its straight streets and lot boundaries, but extends to canals, drains, and railway lines. The straight line was not only 'the offspring of the intentional gaze' as Carter puts it, but is also the instrument par excellence of empire (Carter 1987, p. 222). The straight line rules, okay? The straight line subjects the unruly and recalcitrant wetland to a rationalist, linear logic. The map or plan renders the heights, depths and extension of the land as virtual two-dimensional surface of length and breadth repressing what is below.

In the words of Liam Davison's novel *Soundings*, 'the close-minded straightness of surveyor's lines' (Davison 1993, p. 22) makes possible the roads, dykes and drains which cut across the surface of the land making deep scars on the land. The 'development' of wetlands entails the colonisation of the natural environment which preceded what is ordinarily understood as colonisation by several hundred years, if not by a millennia and a half going back to classical Greek times. Wetlands share with women the dubious privilege of being arguably the first object ever to be colonised. It is hardly surprising then that wetlands and women have been made to share similar qualities in a colonising patriarchal culture, such as passivity and supineness. Both are also thereby the repository and source of patriarchal man's deepest fears and anxieties.

The maps of early Perth enact and trace the history of this colonisation. The 'artificial space of the grid', as Carter puts it, 'brought the "country" into being as country, which created the dialectical boundary between town and non-town. It was through the frame of the town that the bush acquired direction, became a focus of strangeness and an object of desire' (Carter 1987, p. 220), became, in a word, uncanny and the device to overcome uncanniness. Although, as Carter argues, the ultimate effect of what he calls this 'geometrical tendency was to iron out spatial differences, to nullify the strangeness of here and there', this was not performed initially, as he goes on to suggest, as 'a means of translating the country into a place for reliable travelling,' but for safe settling, to render the unhomely homely in the first place. Only then could travel be undertaken. Whilst the elements of the grid certainly rendered, as he puts it, 'the topographical peculiarities of the country [including lakes and swamps] "level" at least in theory,' their initial effect was not to render travelling itself 'an activity independent of place', but to render settling as such, to settle the setting, to make the unhomely homely (ibid. p. 221).

In the intervening years between 1833 and 1838, as the maps of these years show, not only was the grid town extended further out into and over the swamps, but also the unnamed network of lakes and swamps in the earlier map shrank to a set of discrete and independent entities with names in the later map. The naming process served to delimit and contain the boundaries of wetlands and to separate them from the network which once connected them. For de Certeau, 'every power is toponymical and initiates its order of places by naming them', or perhaps more precisely since the Perth lakes and swamps have Aboriginal names by renaming them in the coloniser's terms (de Certeau 1984, p. 130).[11]

After culture colonises nature, after the map colonises space, and the colonists settle in it, the city undertakes rituals of exclusion and repression involving what de Certeau calls 'an excommunication of territorial divinities' (ibid. p. 125). In the case of the wetlands on the Swan Coastal Plain of Western Australia, this entails the excommunication of the Waugal, or 'Rainbow Serpent'. The desanctified territories themselves are regarded as dirt, as matter out of place in Mary Douglas' terms, and, in de Certeau's terms, 'the city founded by utopian and urbanistic discourse . . . must repress all the physical, mental and political pollutions that would compromise it' (ibid. p. 94). Yet the repressed of the city returns in postmodern popular culture in a fascination with the swampy and the slimy, albeit in a sanitised form such as in *Teenage Mutant Ninja Turtles*. Urban life, as de Certeau remarks, 'increasingly permits the re-emergence of the element that the urbanistic project excluded' (ibid. p. 95). Indeed, the element returns in postmodern culture despite all efforts at repressing it.

DECOLONISING WETLANDS

Colonisation is as much about the colonisation of nature as it is about the colonisation of 'the natives', and the colonisation of nature is just as much about the colonisation of 'swamps' as the colonisation of 'the bush'. Indeed, for Fanon, they are one and the same thing:

hostile nature, obstinate and fundamentally rebellious, is in fact represented in the colonies by the bush, by mosquitoes [from swamps], natives and fever [from mosquito bites], and colonisation is a success when all this indocile nature has finally been tamed. Railways across the bush, the draining of swamps and a native population which is non-existent politically and eco-nomically are in fact one and the same thing. (Fanon 1967, p. 201)

The importance, as Deleuze and Guattari put it following Wittfo-gel, of 'large-scale waterworks for an empire' applies not only to aqueducts and dams supplying domestic water to the imperial city but also to the draining and filling of swamps in order to build the city in the first place and develop the agriculture to sustain it (Deleuze and Guattari 1986, p. 21).

The draining of swamps produces the flat and dry surface of the earth on which the straight lines of the railway and the rectilinear grid of the streets and lots of the settler's town can be laid out. It is both a precondition for colonisation of the earth's surface via the construction of the colonial settlement and the development of agriculture and for the colonisation of the earth's depth (like mining) via the reduction of depth to surface in the case of the map or the obliteration of wetlands by draining or filling. Col-onisation of nature gives rise to what Alfred Crosby calls eco-logical imperialism or what John MacKenzie calls simply the empire of nature, our most prized, but unrecognised colony (Crosby 1986; MacKenzie 1988). This natural imperialism, though, is not confined spatially to the former European colonies nor restricted temporally to the nineteenth century and before, but is ongoing everywhere today.

In the era of so-called postcolonialism, it is necessary to ask the question: what process of decolonisation has been carried out in relation to the colonisation of spaces and places, like wetlands, by maps (from which they are absent or on which they are present, reduced to surface and frozen in time), by settlers, and by urban development? Decolonisation will not be fully achieved until all spaces and places are decolonised, not only external, terrestrial and extra-terrestrial space and places, but also internal, corporeal space and places. This includes especially those regions of the human body – the nether regions – associated with the dark and dank regions of the earth – the nether(wet)lands.

What Frederic Jameson calls 'cognitive mapping', or more precisely remapping – even demapping given the complicity between mapping and colonisation – is one route to the decoloni-sation of space and place. Another is corporeal demapping which will re-orientate the subject in relation to his/her own body (rather than regarding, as Jameson does, the body and nature as mean-ingless matter), to the earth, to the regions of both which have been colonised, and to the indigenous owners (Jameson 1984, pp. 89, 90 and 59). This may literally entail walking the streets of Perth with a map of 1838 (such as Figure 8), locating the sites of the lost wetlands and re-locating oneself as embodied subject in relation to them, to the history of their colonisation and that of the dispossession of the indigenous people.

This process of decolonisation of spaces like wetlands and the human body, will be part of the generalised ecology which Felix Guattari envisages: 'ecology should abandon its connotative link with images of a small minority of nature lovers or accredited experts; for the ecology I propose here questions the whole of subjectivity and capitalist power formations' (Guattari 1989, p. 140). Part of that questioning must entail a thorough inter-rogation of the subject's interpellation as body both in relation to space in general and places in particular such as wetlands and in relation to the history of colonial expropriation and imperial exploitation of indigenes and their lands, including wetlands.

NOTES

1. I am grateful to John Fielder for obtaining a copy of the full text of this article.
2. I am grateful to John Fielder for drawing this diary entry to my attention.
3. See Miller 1989, especially Plates 1–6 following p. 114.
4. Dr J. Ferguson, *Colonial Secretaries Office*, vol. 161 (1847), p. 164. I am grateful to John Fielder for obtaining a copy of the full text of this hand-written letter.
5. Ron Blaber, personal communication.
6. I am grateful to Betty Benjamin for drawing this book to my attention.
7. Hugh Webb, personal communication.
8. For further critical discussion of maps, see Harley 1988, pp. 277–312; Harley 1993, pp. 231–47; Hart 1986, pp. 107–16; Huggan 1989, pp. 115–31; and Wood 1992.
9. A similar map of the same year accompanies *Journals of Several Expeditions Made in Western Australia* . . . including one to find the source of the Canning River.
10. In M. Barnard Eldershaw's novel *A House is Built* published in 1929 one of the characters maintains that 'Australia of itself is nothing . . . The country is . . . a *tabula rasa* – a blank sheet'. Quoted by Falkiner, p. 51.
11. See Bekle 1981, pp. 39–41, and Bekle and Gentilli 1993, p. 444 for a list of their Aboriginal and European names.

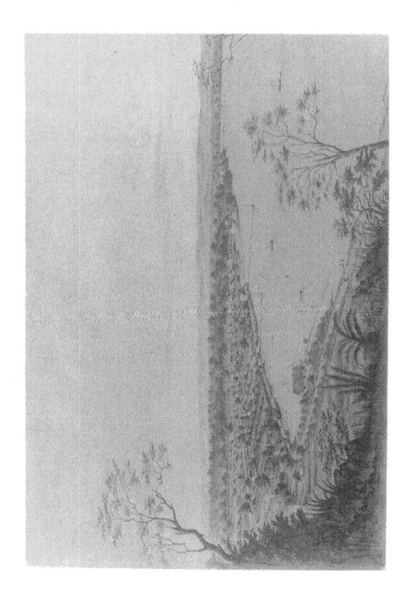

Figure 1   Horace Samson, *Perth 1847*, 1847 watercolour, gouache and pen and ink, 27.5 × 40.3 cm. Collection, Art Gallery of Western Australia, presented by Mr D. Rannard, 1923. Acc 923/W1.

Figure 2   Charles D. Wittenoom, *Sketch of the Town of Perth from Perth Water, Western Australia,* 1839 line lithograph, 15.8 × 24.2 cm (image). Collection, Art Gallery of Western Australia, transferred from the Public Library, 1916. Acc 916/Q2.

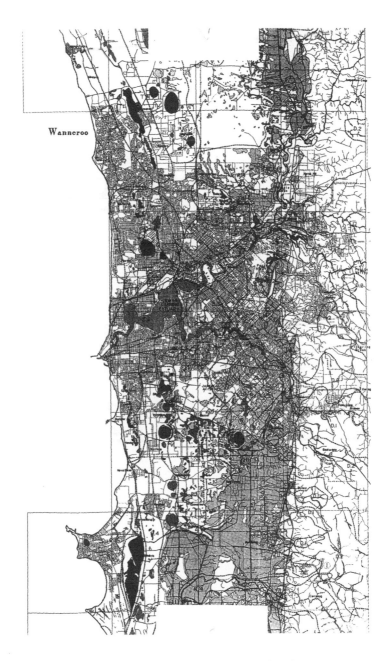

Figure 3 Detail showing Perth metropolitan area taken from geomorphic
wetland mapping conducted for the Water Authority of Western Australia,
Perth: A City of Wetlands, 1989–90. .

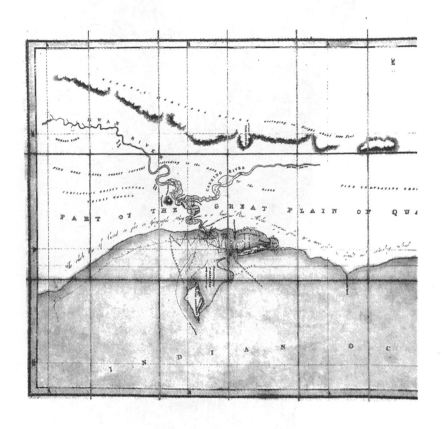

Figure 4   Detail from Captain James Stirling, RN, *Chart of Part of the Western Coast of New Holland . . .*, 1827. Collection, Battye Library of Western Australian History. Acc 27C.

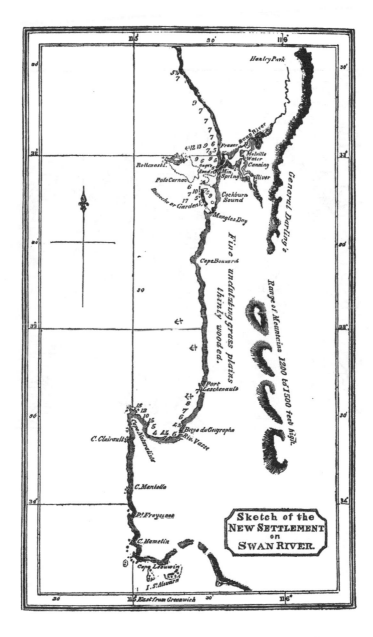

Figure 5 *Sketch of the New Settlement on Swan River, Quarterly Review,*
vol. 39 (January and April 1829), p. 323. Collection, Battye Library of
Western Australian History.

Figure 6   Detail from J. Arrowsmith, *Discoveries in Western Australia . . .*,
1833. Collection, Battye Library of Western Australian History. Acc 1185C.

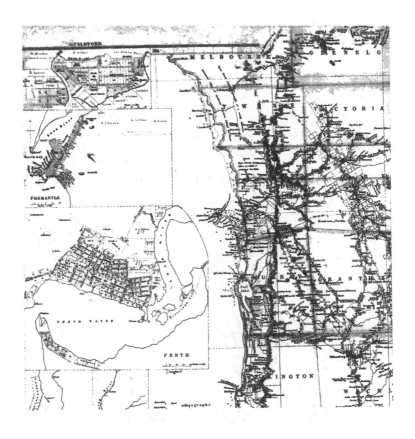

Figure 7 Detail from J. Arrowsmith, *The Colony of Western Australia . . .*, 1838. Collection, Battye Library of Western Australian History. Acc 90C.

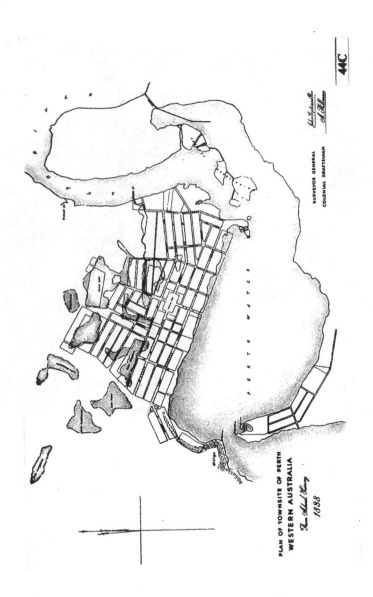

Figure 8   A. Hillman and John Septimus Roe, *Plan of Townsite of Perth,
Western Australia* . . ., 1838. Collection, Battye Library of Western Australian
History. Acc 44C.

4

The World/Womb as Wetland, the Modern City as Cultural Symptom and the Postmodern City as Swamp

The modern city is made possible by maps and by mapping. The map makes an inscription, a mark on the surface of paper, parchment, or drawing film. The modern city itself, though, is also an inscription, a set of marks on the surface of the earth, including, most particularly, wetlands. The mark on the map of city allotments makes possible and is aligned culturally with the city lots marked on the earth's surface by survey pegs and lines, and later by buildings and roads. In the process the earth is reduced to surface, its contours, its heights and depths, reduced to a virtual two-dimensional plane, 'the space of property which coincides with the "dead surface",' as Derrida puts it (Derrida 1982, p. 331). The dead surface of property has often been inscribed over the living depths of wetlands.

Using Freud's theory of the symptom as inscribed on the surface of the patient's body and applying it to the city as a cultural symptom of what Aldo Leopold calls 'a land pathology' (Leopold (1935) 1992, pp. 212–17), or more precisely a psychopathology, inscribed on the surface of the earth (hence a psychogeopathology), I read the modern city, and the project of modernity, as symptomatic of what could be called a wetland pathology, or more elaborately, a psychogeopathology of the will to fill wetlands. By 'speaking the symptom' Freud believed it would disappear. But as Zoë Sofoulis argues, this simply does not occur.[1]

Even though the city will not disappear just because it is spoken that it is symptomatic of a repression of the swamp, speaking the city as a symptom of psychogeopathology is a kind of cultural 'talking cure' which raises the swamp as (a figure of) the cultural unconscious, indeed the cultural repressed, to consciousness and de-pathologises it. Without that 'consciousness raising', without that 'speaking the symptom', wetlands conservation and rehabilitation will not be addressing the cultural conditions of wetlands colonisation and destruction, but will merely be attempting to block a tide of culturally habituated opinion going against conservation and rehabilitation of wetlands.

CITY SYMPTOMS

If the city is a symptom, it has a repressed. Beneath the surface of the city is the repressed of the city. Henri Lefebvre has suggested: 'if it turned out . . . that every society, and particularly . . . the city, had an undergroupd and repressed life, and hence an 'unconscious' of its own, there can be no doubt that interest in psychoanalysis, at present on the decline, would get a new lease on life' (Lefebvre 1991, p. 36).[2] All sewered cities have a repressed in the sewerage system beneath them. Some cities have also repressed the swamps on which they were built.

The swamps beneath, or before, the city are repressed spatially and historically. Reading old maps of the city in terms of the way in which they repress swamps enables that history to come to light. The map was a powerful instrument of colonisation in the hands of early explorers, surveyors and settlers but is now a significant symptom of their will to fill wetlands. The historical repressed of the city, the swampy history which the city would like to forget it had, is decolonised in the psychoanalytic reading of the map. Psychoanalysis can be given a new lease of life by raising the unconscious, indeed repressed, of the city to consciousness and by a psychoanalytic ecology which would undertake a cultural talking cure of the psychogeopathology of the will to fill, or aestheticise, or create artificial, wetlands.

The possibility of a psychoanalytic cultural studies has repeatedly been entertained in order to address the cultural unconscious so often neglected by mainstream cultural studies. It would not simply be diagnostic of the cultural (and environmental) unconscious but as a talking cure of the cultural symptom would undermine the conditions that make possible that symptom's pathological repetition. So, although the city will not disappear as symptom of the psychogeopathology of the will to find, the wetland as a psychogeopathological category of the horrific and the monstrous may.

If the modern city is symptomatic of the psychogeopathology of a will to fill wetlands, and if the swamp is (a trope for) the cultural unconscious, more particularly its repressed, then the repressed returns in cultural dreaming, jokes and slips of the tongue. This return of the repressed can be seen in postmodern popular culture's fascination with and horror of the dark and dank underside of the city. This surfaces (excuse the pun), albeit in a sanitised form, in *Teenage Mutant Ninja Turtles*.

In this chapter I trace the return of the swamp as the return of the repressed of the city in Richard Jefferies' early work of science fiction *After London* and in J. G. Ballard's second novel *The Drowned World*. The modern city is often built on filled or drained swamps. Visions of the postmodern city often involve the return of the repressed. The gender politics of these concerns are extensively traced in Joseph Conrad's first short story 'The Lagoon' and in Janet Frame's short story of the same title in her first published collection.

If inscriptive writing is colonising, especially via map-making, then speaking the symptom and undertaking a cultural 'talking cure' would be decolonising rather than merely deconstructive. Yet to restore the symptom to full speech as Freud does, to try and make the symptom disappear, would be tantamount to recolonising the already colonised. It would be to re-assert the colonial and colonising power of full speach in the very act of deciphering its written symptoms. Colonisation writes on surfaces, not only on the surface of the 'native's' and the convict's body, but also on the surface of the earth via the city and the map.[3] Decolonisation reads these surfaces for their symptomatic manifestation of the cultural repressed and the economic and political oppressed. Speaking the symptom of the city and the map without restoring it to full speech decolonises the wetland.

In Lectures 17 and 18 of the *Introductory Lectures*, Freud theorises the symptom as having a sense inscribed on the surface of the patient's body.[4] Likewise more recently Lacan saw a symptom as 'a metaphor in which flesh or function is taken as a signifying element' (Lacan 1977, p. 110). If symptoms are inscribed on surfaces, as Freud contends, then they are a sort of writing. Conversely, Derrida argues that 'writing belongs to the order and exteriority of the symptom' (Derrida 1982, p. 110). Extrapolating more generally from Freud, Derrida and Lacan, and taking further Deleuze and Guattari's suggestion that a surface of inscription is necessary for social reproduction and that the earth is one such surface, a symptom can be inscribed on any surface – bodies, the earth, pages, – and a symptom is cultural irrespective of whether it is inscribed on an individual body, for example, anorexia nervosa, or on the earth, for example, cities in or on former swamps. These symptoms are constructed from unconscious processes and are symptomatic of what Zoë Sofoulis calls an arrested or blocked process[5] which dam the flows of the libido and stem the tides of desire.

The process of arresting or blocking is pathological. A symptom for Freud 'actually denotes the presence of some pathological process' (Freud 1979, p. 237). Manifested in the symptom of the city and the map, and repressed by them, is the psychogeopathology of the colonisation of nature in general and wetlands in particular. The marks which colonialism leaves on its maps and the marks its cities leave on the earth are symptomatic of a psychogeopathology. A pathology is indicated in general, as Leopold points out, by 'self-accelerating rather than self-compensating departures from normal functioning' (Leopold (1935) 1992, p. 217). In the case of a land pathology in particular this self-acceleration occurs in the 'collective organism of land and society' (ibid). The growth of cities has certainly been self-accelerating over the last one hundred years.[6] Cities have departed from the normal, indigenous functioning of the land/society organism, not least in the relationship with wetlands which had obtained, in Australia for example, for 50,000 years before.[7]

THE SWAMP STRIKES BACK

If the sewered city built on a swamp has a repressed, then the repressed returns in cultural dreams, puns and slips of the tongue about the sewers beneath and the swamps before or beneath or outside the city. One such city is London, the setting for Richard Jefferies' post-apocalyptic dystopian novel *After London*, first published in 1885 (Jefferies (1885) 1939). The novel has been described by Raymond Williams as 'a powerful but acrid vision of the metropolis reclaimed by the swamp' (Williams 1973, p. 196). Instead of the city reclaiming the swamp (as it had done), the swamp reclaims the city. Much of the east end of London was built on drained marshes; in the novel the whole city is retaken by them and so history reversed. This can be seen as the playing out of the return of the repressed on a number of levels, or on a number of congruent sites – geographical, psychological, and so on. It can also be seen as a kind of revenge of history in which the lost marshes of London return to destroy it. This theme of the destruction of London by water is taken up later and developed more extensively by J. G. Ballard in *The Drowned World* (Ballard, 1983).[8]

In *After London* much of England is a shallow lake, a wetland. At its eastern extremity the lake narrows and 'finally is lost in the vast marshes which cover the site of ancient London' (Jefferies (1885) 1939 p. 32). The Thames as such disappears, though it had already 'become partially choked from the cloacae of the ancient city' (ibid. p. 32). As a result of the inundation, 'the mighty metropolis was soon overthrown' and 'the low-lying parts of the mighty city of London became swamps' (ibid. p. 33) complete with dead black water. London is

a vast stagnant swamp, which no man dare enter, since death would be his inevitable fate. There exhales from this oozy mass so fatal a vapour that no animal can endure it. The black water bears a greenish-brown floating scum, which for ever bubbles up from the putrid mud of the bottom. When the wind collects the miasma, and, as it were, presses it together, it becomes visible as a low cloud which hangs over the place. (ibid. p. 33)

This cloud of compressed miasma is reminiscent of Dickens' mudfog.

Rising from the dead black water of the swamp, the miasmic cloud sometimes obscures the sun, generally making the whole scene depressing beneath the now black sun of depression and melancholy. The dead black waters of the swamp have, though and in turn, arisen from even further below: 'all the rottenness of a thousand years and of many hundred millions of human beings is there festering under the stagnant water, which has sunk down

into and penetrated the earth, and floated up to the surface the
contents of the buried cloacae' (ibid. p. 33). The repressed of the
modern bourgeois city, the excrement, the waste, previously
hidden in its sewers below the city, out of sight and out of mind,
here literally returns and surfaces to flood the city.

The use of the word 'cloacae' is interesting because 'cloaca' is,
according to the *Concise Oxford Dictionary*, sewer, excremental
cavity in some animals and gathering place of moral evil, in other
words, a sump of iniquity. More generally it simply means a single
opening for excretory and reproductive functions. Ballard uses the
word repeatedly in *The Drowned World*. The underside of the city
is troped as a kind of animal body with its sewers which are its
single opening for its digestive, genital (and so reproductive) and
excretory functions. The city digests, defecates and procreates
using the same organ, its sewers. The city excretes and reproduces
through the same hole. The underside of the city is thereby
feminised. As Marie Bonaparte argues 'a woman, in fact, possesses
a cloaca, divided by the recto-vaginal septum into the anus and
the specifically feminine vagina, the gateway to the additional
structure of the maternal apparatus' (Bonaparte 1953, p. 175). The
horror of cloacae is inflected with a misogynist horror of women's
bodies, especially the mother's.

The cloaca is the obscene of the city in a number of senses.
Lefebvre notes 'the general fact that walls, enclosures and facades
serve to define both a *scene and an obscene area to which
everything that cannot or may not happen on the scene is
relegated: whatever is inadmissible, be it malefic or forbidden,
thus has its own hidden space on the near or the far side of a
frontier'* (Lefebvre 1991, p. 36). This hidden space and obscene
area is the sewers beneath the city and the swamps which the city
may have excluded by its coming and on which it is built. The
frontier which was marked out in two dimensions on the surface
of the earth and the map (and which was overcome by the city
colonising both) has been transformed into the frontier between
surface and depth, between the dimensions of length and breadth,
or width, and the dimension of depth.

The frontier on the surface has been construed as a defining
boundary between civilisation and wilderness. It is also important
for the constitution of private property. A defining characteristic

of (private) property for Lefebvre, as well as of 'the position in space of a town, nation or nation state, is a closed frontier' (ibid. p. 176). Yet the closed frontier does not operate only on the surface between wilderness and civilisation, town and country, private and public property but also between surface and depth, city and swamp. The re-assertion of the sewers and swamps in texts like Jefferies' is a kind of return of the repressed of the city and the map in which the city is turned belly side up to show its rank and reeking underbelly.

This re-assertion not only plays out urban fears but also imperial aspirations. The city in *After London* has, according to U. C. Knoepflmacher, 'quite literally disappeared – in its place there is a dark primeval swamp penetrated only by the most intrepid explorers' (Knoepflmacher 1973, p. 535), just like the swamps of Africa in British imperialist adventure stories such as *African Queen* and *She* and the 'swamps' of nineteenth-century working-class slums penetrated by sanitary inspectors and reformers. The bourgeois bifurcation of surface and depth reduces the depths of the Body of the Mother to surface and represses the depths of the city. The surface of the city and the swamp is then penetrated by phallic heroes who invaginate that surface.

The Return of the Colonial and Uterine Repressed

Jefferies' dystopian, apocalyptic vision seems to have been consciously taken up and developed by J. G. Ballard in his second novel, *The Drowned World*, first published in 1963 in which a flooded London has been transformed into a tropical lagoon. But unlike Jefferies' London in which 'the very largest of the buildings fell in' (Jefferies (1885) 1939, p. 33), in Ballard's London some of its skyscrapers have been left with their tops sticking up out of the water. The world below has been drowned, but the skyscrapers continue to serve a vestigial function for which they were designed – to transcend the world below, to sublimate the base matter of the earth below (the slime) into the supreme heights of abstraction (the sublime) repressing what is below (see Chapter 2, Figure 1). Both *After London* and *The Drowned World* are concerned with the return of the spatial repressed (the tropical colonies and the uterus), temporal repressed (the uterine and colonial pasts) and psychological repressed (memories of the womb and the mother's body). Indeed, for *The Drowned World* there is a homology between all three.

Unlike Jefferies' vision in which London is flooded as a return of the drained marshes of the Thames, *The Drowned World* has the previously drained world, the temperate dry lands, of London flooded in a return of the repressed and colonised tropical 'other', the tropical wet lands. In *After London* a return of the repressed within – the endocolonised marshes of temperate England – takes place; in *The Drowned World*, what occurs is a return of the repressed without, the exocolonised lagoons of the tropics. The boundary between the tropical and the temperate ceases to exist with the former's region and climate having taken over the latter's. London, and the whole of England for that matter, is now located in 'the European lagoons' (Ballard (1963) 1983, p. 12).

The inhabitants of London become so many latter day colonists of the tropical 'at home'. The lagoon which was not 'at home', but was colonised in the tropical, has come home; the wetlands which were not like those 'at home' have come home and made home their own; the unhomely has invaded and conquered the homely. In this return of the colonial repressed the official job of Kerans, the central character of the novel, is to serve a vestigial colonial function of cartographer by 'mapping the shifting keys and harbours and evacuating the last inhabitants . . . living on in the sinking cities' (ibid. p. 12). As the story progresses, however, Kerans is transformed from imperialist cartographer to wetlands conservationist. I have already discussed the difficulty of mapping wetlands spatially because of their changing status temporally. By an etymological play on 'lagoon' and 'lacuna', both of which come from the Latin '*lacuna*' meaning pool, Kerans' job is made more difficult because a lacuna (and a lagoon on a map) is 'a hiatus, blank, missing part, a gap, an empty space, spot, or cavity', like a cloaca.[9] For the cartographer the lagoon, the slime (and the wetland more generally) is an impossible abject because of its changing shape.

Yet Kerans undergoes a transformation in his perception of the wet land. Initially, for him 'the lagoon was nothing more than a garbage-filled swamp' (Ballard (1963) 1983, p. 13). Culturally, and spatially, the two are not so far apart as wetlands have often been the site of 'sanitary land-fill'.[10] The description of the swamp as 'nothing more . . .' upholds the western tradition of reductive, pejorative markers being assigned to wetlands. Riggs concurs with Kerans's summation when he pronounces that 'the whole place is nothing but a confounded zoo' (Ballard (1963) 1983, p. 17) which is like Jefferies' reduction of the city to the animal. Yet unlike Jefferies' swamp, which is unhealthy in accordance with the prevailing miasmatic theory of disease of his time, Ballard's lagoon simply smells bad: 'Kerans felt the terrible stench of the water-line, the sweet compacted smells of dead vegetation and rotting animal carcasses' (ibid. p. 13). Kerans feels, rather than smells, the stench. Perhaps he would rather feel the stench than smell it. The sensation of touch stays on the outside whereas the sensation of smell penetrates inside. For Kerans 'a thick cloacal stench exuded from the silt flat' (ibid. p. 61).

The return of the repressed, of the bad smells of the swamp, of the colonial other, also involves the return to the past, both the past of the human species and the individual past, in a reversal of the Freudian tenet following Haeckel that individual development repeats the development of the species (ontogenesis repeats phylogenesis). In Ballard's novel the apocalypse of the world drowning reverses phylogenesis in a return to the primeval slime, or more precisely to 'impenetrable Mato Grossos sometimes three hundred feet high . . . a nightmare world of competing organic forms returning rapidly to their Paleozoic past . . .' (ibid. p. 19).

Ontogenesis is also reversed for Kerans in a return to amniotic fluid, a prospect which fills him initially with horror when he wonders whether 'perhaps these sunken lagoons simply remind me of the drowned world of my uterine childhood – if so, the best thing is to leave straight away' (ibid. p. 28). The lagoon, perhaps of all wetlands, is markedly, though naturally, enclosed. On a continuum of wetland types from the most earthbound to the least, the lagoon is the most wet of the wetlands. It is also separated or closed off from the sea by low sandbanks or by an atoll, a small sea within a larger. The lagoon is a kind of natural vessel made, like all wetlands, of earth and water, but with a thinner shell, a rim and even a lip. The drowned world is referred to as 'the black bowl of the lagoon' (ibid. p. 47, see also p. 70). No doubt the black bowl contains black water.

This bowl of black water is archetypally the womb. In his archetypal and misogynist study of the Great Mother, Erich Neumann has suggested that: 'the natural elements that are essentially connected with vessel symbolism include both earth and water. This containing water is the primordial womb of life, from which in innumerable myths, life is born. It is the water 'below', the water of the depths, ground water and ocean, lake and pond' (Neumann 1955, p. 47). And lagoon. These waters are highly fertile. In particular, the littoral, the intertidal zone, the estuarine and lagunine, are some of the most productive eocosystems on the planet, up to twenty times richer in wildlife than the open ocean (Earll 1991, p. 161).

For Kerans the drowned worlds of the phylogenetic and ontogenetic, of the exterior and interior, past and present become indistinguishable. The uterine past of the individual and the paleozoic past of the human species have been repressed. Their joint return is represented in the iconography of the drowned world itself and portrayed in 'one of Max Ernst's self-devouring phantasmagoric jungles [which] screamed silently to itself, like the sump of some insane unconscious' (Ballard (1963) 1983, p. 29). The unconscious is even more strongly figured in aquatic terms in the reference to 'the submerged levels below his consciousness' (ibid. p. 83). *The Drowned World* seems to have an equally strong intertextual relationship with Ernst's painting, 'Europe After the Rain', with its nightmarish vision of what Robert Hughes calls 'a panorama of a fungoid landscape seen as though in the aftermath of an annihilating biblical deluge' (Hughes 1980, p. 255) as it does with *After London* in its apocalyptic vision of the city as a stagnant swamp of dead black water after the flood.

A powerful gender politics and psychodynamics is enacted here between the swamp, the unconscious and the feminine. The swamp has been used in patriarchy as a figure for the unconscious and for female sexuality. Erich Neumann argues that

the female powers dwell not only in ponds, springs, streams and swamps but also in the earth, in mountains, hills, cliffs, and – along with the dead and unborn – in the underworld. And above all, the mixture of the elements of water and earth is primordially feminine; it is the swamp, the fertile muck, in whose uroboric nature the water may equally well be experienced as male and engendering or as female and birthgiving. (Neumann 1955, p. 260)

The lagoon is the most watery mixture of earth and water, the wetland with the least possible amount of land, and so perhaps the wetland to be most strongly feminised in the western tradition.

To complete the triangle of troping between the unconscious, the feminine and the swamp, patriarchy has gendered the unconscious as feminine and the conscious as masculine (ibid. pp. 147–8). Luce Irigaray has wondered 'whether certain properties attributed to the unconscious may not, in part, be ascribed to the female sex, which is censured by the logic of consciousness. Whether the feminine has an unconscious or whether it is the unconscious' (Irigaray 1985, p. 73).[11] One of the properties attributed to the unconscious which has also been ascribed to the female sex is fluidity. Indeed, Irigaray elsewhere maintains that 'historically the properties of fluids have been abandoned to the feminine' (ibid. p. 116).[12] Yet it is only certain properties of certain fluids which have been thus abandoned to the feminine by patriarchy. The life-giving properties of fecund and nurturing fluids of the mother's body have been regarded as static and stagnant whereas the death-dealing properties of clear and clean fluids have been seen as dynamic and flowing, and have been appropriated by the masculine (see Chapter 2, Figure 1).[13]

Rivers have been masculinised and swamps feminised, the latter most powerfully in the organising principle of *The Drowned World*, the wetland as womb, though the novel develops this trope from the modality of horror to that of fascination. Rather than enacting a typically masculine repudiation of his birth, Kerans becomes fascinated with his uterine experience though not with his mother per se. He shows no interest in his mother as a person. His fascination with his uterine experience is abstracted from the body in which it occurred. Kerans' mother is conspicuous by her absence from his memory and his curiosity. She is a hiatus, blank, missing part, a gap, an empty space, in short, a lacuna in the story of his life, a lagoon in the map of his life.

Perhaps not surprisingly then, Keran's attitude to Beatrice, the only female character in the novel, is inflected with misogyny. The choice of name is rather ironic given Dante's fascination with his own Beatrice and also with the slimy in Canto 7 of the *Inferno*. The wetland hell of *The Drowned World* could be read as a postmodern version of Dante's medieval vision of a Stygian marsh with its souls of the sullen stuck in slime. For Kerans, though, the Beatrice of *The Drowned World* is not the beatific Beatrice of Dante's *Divine Comedy*, but 'Pandora with her killing mouth and witch's box of desires and frustrations, unpredictably opening and shutting the lid' (Ballard (1963) 1983, p. 31). Beatrice is represented as an automata, that ultimate figure of the *unheimlich* or uncanny for Freud, with a devouring, mechanical and oral-sadistic mouth.

In an essentialist gesture, for Neumann, 'the mouth as rending, devouring symbol of aggression is characteristic of the dangerous negative elementary character of the Feminine' (Neumann 1955, pp. 122–3). Or more precisely, the fear of being eaten by the mother in an act of symmetrical reciprocity characterises the oral-sadistic phase interposed between what Freud calls the 'primitive oral organisation – the fear of being eaten up . . . by the father' and the 'anal-sadistic phase – the wish to be eaten by the father' (Freud 1984, p. 149). The fear of symmetry and fairness being exercised back against oneself characterises the oral-sadistic organisation (Giblett 1993, pp. 541–59).

As an enclosed, moist space with a single opening the mouth – especially before teeth appear – is associated with the womb in the anal phase where both are seen as passive organs. For Bonaparte, 'in the earlier anal stage, it was the cloaca that appeared as the erotogenic organ. Then, as throughout life, the cloaca is a passive organ as largely is the mouth before the teeth appear, in the primary stage' (Bonaparte 1953, p. 24). But when teeth appear the mouth, and the cloaca by previous association, can become active, even sadistic. The functions of both are inverted with the mouth becoming the organ of (re)production as the cloaca was (and still is) and the cloaca the site of consumption and destruction as the mouth had been. For Neumann:

the positive femininity of the womb appears as a mouth; that is why 'lips' are attributed to the female genitals, and on the basis of this positive symbolic equation the mouth, as 'upper womb', is the birthplace of the breath and the word, the Logos. Similarly, the destructive side of the Feminine, the destructive and deadly womb, appears most frequently in the archetype form of a mouth bristling with teeth . . . the *vagina dentata*. (Neumann 1955, p. 168)[14]

Beatrice's 'killing mouth' like Pandora's box is a *vagina dentata* for Kerans and Kerans' fascination with his uterine past is displaced onto and re-enacted with her as a fear of ingestion and entrapment.

For whom or for what, though, is the womb, in Neumann's terms, destructive and deathly? Only for the phallic man and the phallus, as Neumann goes on to suggest, albeit in misogynist terms: 'the Feminine – as avid womb – attracts the male and kills the phallus within itself in order to achieve satisfaction and fecundation' (ibid. p. 171). Yet by phantasising the *vagina dentata* as an orally sadistic organ, the phallic man demonstrates his impotence. As Marie Bonaparte suggests 'the cloaca . . . will never be dentated except in the phantasies of impotent men who fear it, as such, in woman' (Bonaparte 1953, p. 24). For phallic men the vagina is a sperm spittoon and the womb, as Irigaray puts it, 'is variously phantasized as a devouring mouth, as a sewer in which anal and urethral waste is poured, as a threat to the phallus or, at best is as a reproductive organ' (Irigaray 1993, p. 16) divested of all other qualities – an incubator on legs. Similarly, the wetland is a site for sanitary land-fill and phantasised as a devouring mouth, as a sewer in which anal and toxic waste is poured, as an engulfing threat to phallic power, or, at most, a reproductive organ – a fish hatchery, a wombland.

Women's desire, Irigaray argues, is 'often interpreted, and feared, as a sort of insatiable hunger, a voracity, that will swallow you whole. Whereas it really involves a different economy more than anything else, one that upsets the linearity of a project, undermines the goal-object of a desire, diffuses the polarisation toward a single pleasure, disconcerts fidelity to a single discourse' (Irigaray 1985, pp. 29–30). The fluidity of women's desire threatens to undermine the shores, banks, dykes, of phallic men's objects, private property, things, just as the sea and the swamp threaten to absorb 'developments' built between them.[15]

For Neumann, it is only 'the hero' who can escape this fate of re-absorption: 'the womb, the gate, the gullet . . . actively swallows, devours, rends and kills. Its sucking power is mythologically symbolised by its lure and attraction for man, for life and consciousness and the individual male, who can evade it only if he is a hero, and even then not always' (Neumann 1955, p. 171). For the phallic man the womb is an orally sadistic hell sucking at him, like the slime which sucks at Sartre, which he can only escape if he is prepared to descend heroically into it and find his way through it. In fact, heroism is constituted in patriarchy by the ability to survive the descent into the underworld, or 'dark continent', or 'black water', of female sexuality as I will try to show in a later chapter. Kerans is a hero who does literally descend into the underworld of the drowned world, or womb, of London and not only survives but undergoes a transformation, or kind of rebirth.

Kerans' fascination with his uterine past and its homologous relationship with the Paleozoic past of the human race are confirmed by Bodkin: 'just as psychoanalysis reconstructs the original traumatic situation in order to release the repressed material, so we are now being plunged back into the archae-opsychic past, uncovering the ancient taboos and drives that have been dormant for epochs' (Ballard (1963) 1983, pp. 43–4). Plunged back, in other words, into the repressed world of primeval slime and the swamp as a kind of recapitulation of birth in reverse and in the process raising the unconscious to consciousness in a successful 'talking cure'.

For Freud, 'the task of a psychoanalytic treatment . . . is to make conscious everything that is pathogenically unconscious . . . and to fill up all the gaps in the patient's memory, to remove his amnesias' (Freud 1973, p. 323). In other words, to fill up, and so fill in, the lacunae (lagoon) in the surfaces of the patient's memory about the womb/wetlands as the place from whence we came and as the underside of the city and thus to repeat colonising inscrip-tive writing. The task of a psychoanalytic ecology, on the other hand, would be to read the absences in the maps of the wetlands/womb as symptomatic of a psychogeopathology of the will to fill wetlands, and so decolonise inscriptive writing.

Bodkin goes on to suggest that 'the uterine odyssey of the growing foetus recapitulates the entire evolutionary past' (Ballard (1963) 1983,p. 44), or in other words, ontogenesis recapitulates phylogenesis from primeval slime to postmodern times as a heroic journey. Individually, and ontogenetically, for Bodkin, 'we really remember these swamps and lagoons' (ibid. p. 48). The lacunae in memory and the lagoon on the map are filled up. The lagunine apocalypse of a decaying Europe, however, recapitulates in reverse the entire evolutionary past from postmodern times back to primeval slime to arrive at 'the very junction where we stand now on the shores of this lagoon, between the Paleozoic and Triassic Eras' (ibid. p. 44). This is both a form of time-travel in which the earth (and its history) is traced temporally (including its wetlands unlike the spatial tracing of the earth (and its geography) on maps which invariably do not) and of what Bodkin graces with the high-sounding name of 'the Psychology of Total Equivalents' or 'Neuronics':

I am convinced that as we move back through geophysical time so we re-enter the amnionic [sic] corridor and move back through spinal and archaeopsychic time, recollecting in our unconscious minds the landscape of each epoch, each with a distinct geological terrain, its own unique flora and fauna, as recognisable to anyone else as they would be to a traveller in a Wellsian time-machine. Except that this is no scenic railway, but a total re-orientation of the personality. If we let these buried phantoms master us as they re-appear we'll be swept back helplessly in the flood-tide like pieces of flotsam. (ibid. pp. 44–5)

Like Kerans, Bodkin fears being plunged or swept back in metaphors of fluidity into the past, whether it be the uterine or Paleozoic past or both. It is matter of master or be mastered, or more precisely mistressed. Even though the tropical and the orient have won out climatically and geographically over the temperate and the occident, the battle still rages psychologically between these sites in the individual and collective mind and body. The colonies have struck back and triumphed externally, but internally the battle still goes on.

Much of that battle is fought out, and lost, as would be expected following Freud's work, in the domain of dreams. Kerans has a dream in which he steps out into the lake 'whose waters now

seemed an extension of his own bloodstream. As the dull pound-
ing rose, he felt the barriers which divided his own cells from the
surrounding medium dissolving, and he swam forwards, spreading outwards across the black thudding water' (ibid. p. 71). This
dream of the dissolution of Keran's ego in black water is a reversal
of the dreams which symbolise the ego's formation. For Jacques
Lacan 'the formation of the *I* is symbolised in dreams by a fortress,
or a stadium – its inner arena and enclosure, surrounded by
marshes and rubbish tips, dividing it into two opposed fields of
contest where the subject flounders in quest of the lofty, remote
inner castle whose forms . . . symbolise the id in a quite startling
way' (Lacan 1977, p. 5).[16]

The exchanges, as David Miller puts it, in the swamp, as well as
in the lagoon (and other wetlands more generally), 'between skin
and enveloping matter . . . all initiate the dissolution of the self'
(Miller 1989, p. 208). In Kerans' dream he leaves the security of
what Antony Easthope calls the castle of the self (Easthope 1986,
pp. 35–44) and flounders around in the black waters and garbage
dump of London's lagoons. This dream is a recurring one not just
involving the dissolution of the boundaries of his body and his
ego, but also a return to his amniotic past. The lake becomes 'the
warm amnionic [*sic*] jelly through which he swam in his dreams'
(Ballard (1963) 1983, p. 99). Yet this womb associated with life is
also ambivalently, and misogynistically, figured as a tomb associated with death.

All Kerans' fears of ingestion and dissolution are re-enacted
when, in the ninth chapter entitled, 'The Pool of Thanatos', he
dives into 'a tank filled with warm, glutinous jelly that clamped
itself to his calves and thighs like the foetid embrace of some
gigantic protozoan monster' (ibid. p. 104). Given that he is diving
into water troped as amniotic fluid, the body of his mother is, by
implication, 'a protozoan monster' and his uterine childhood a
fetid, (s)mothering embrace from which he has escaped and to
which he does not want to return. Kerans is part of that more
general patriarchal process, and project, whereby, as Irigaray puts
it, 'we have turned the mother into the devouring monster'
(Irigaray 1993, p. 15). When he is below he is asked 'How's the
grey sweet mother of us all?' to which he replies 'Feels like home'

(Ballard (1963) 1983, p. 105), but an unhomely or *unheimlich* home which is uncanny.

Kerans descends further into this uncanny home by entering an auditorium 'with its blurred walls cloaked with silt [which] rose up above him like a huge velvet-upholstered womb in a surrealist nightmare' (ibid. p. 108). Kerans almost dies from diving too deep, but in the process experiences a kind of rebirth and the realisation of his own dream of dissolution of his ego and body boundaries:

as the spotlight flared across the domed ceiling, illuminating the huge vacant womb for the last time, Kerans felt the warm blood-filled nausea of the chamber flood in upon him. He lay back . . . the soothing pressure of the water penetrating his suit so that the barriers between his own private bloodstream and that of the giant amnion seemed no longer to exist. The deep cradle of silt carried him gently like an immense placenta, infinitely softer than any bed he had ever known. (ibid. p. 110)

When Kerans' dream is actualised, it is not as horrific as he feared it might have been, and the embracing, entangling protozoan mother-monster turns out to be an immense, soft placenta with living black waters. Kerans' view of wetlands changes from oral-sadistic, consuming and death-dealing *vagina dentata* to corporeal-pleasurable, (re)producing and life-giving placenta. The silt is no longer horrific and nightmarish, but comforting and nurturing. The velvet-upholstered womb in a surrealist nightmare becomes a deep cradle in a pleasant daydream. The auditorium becomes transformed in recollection into 'the great womb-chamber of the planetarium' (ibid. p. 112), an exterior archi-tectural analogue, or homologue, of interior bodily space. Kerans is likewise transformed into what Donna Haraway calls (and calls herself) 'a planetary foetus . . . gestating in the amniotic effluvia of terminal industrialism' (Haraway 1992, p. 296), parasitic on placental wetlands past and present.

So when Strangman, the villain of the story, undertakes the project of pumping dry a section of London, 'turning the once limpid beauty of the underwater city into a drained and festering sewer', Kerans is 'unable to accept the logic of the rebirth before him' (Ballard (1963) 1983 p. 121), a rebirth unlike the one which he has just experienced in that instead of returning to the deep, to the amniotic, as he himself did, part of the lagoon is pumped dry in an attempt to return London to its pre-drowned state. For Beatrice, the drained section of London is 'like some imaginary city of Hell' (ibid. p. 123), a necropolis, in fact (ibid. p. 151). This hell is Dantesque, or at least like the fifth circle of Dante's *Inferno* set in the slime of the Stygian marsh, especially with 'the black slime oozing' out of the pores of the protozoan mother-monster: 'everything was covered with a fine coating of silt, smothering whatever grace and character had once distinguished the streets, so that the entire city seemed to Kerans to have been resurrected from its own sewers. . . The once translucent threshold of the womb had vanished, its place taken by the gateway to a sewer' (ibid. pp. 126–7). Vagina and anus are here united in the cloaca of the city-mother (ibid. pp. 53 and 61) as they are for Marie Bonaparte in women more generally. Drowned London had reverted from modern city to primeval swamp and now redrained London reverts from primeval swamp to postmodern slime and sewer covered in silt.

In accordance with the fundamental structuring principle of the novel whereby inner and outer are analogous, if not homologous, the drained lagoon affects Kerans profoundly and 'he began to sink rapidly into a state of dulled inertia, from which he tried helplessly to rouse himself. Dimly he realised that the lagoon had represented a complex of neuronic needs that were impossible to satisfy by any other means' (ibid. p. 129). Kerans experiences a sort of post-natal depression, not as mother, but as someone who has just been (re-)born. As a result, he becomes a kind of wetlands conservationist, though hardly an ecofeminist, who is sensitive to the homology between the inner and outer landscapes and aware of the connections between the depths and the surfaces, the repressed and the expressed, the past and the present. Although he is now fascinated with and not horrified by the wetland as womb, he is still not interested in his mother, and the Body of the Mother. The mother (in two senses) is absent from his story, though it/she is virtually present, or has left traces, in the references to the protozoan monster and black water of the lagoons. Despite his transformation from imperialist cartographer, Kerans is still a hero from whose life, whose story, his mother is largely absent. Kerans extends his wetlands conservationism to becoming an eco-saboteur who blows up the dykes which hold the water back and then heads south 'a second Adam searching for the forgotten paradises of the reborn sun' (ibid. p. 175).

Ballard's work is important as an attempt to tell the story of the imperial city which colonised wetlands at 'home and abroad' and further to think the unthinkable future albeit as a repetition of the past in the return of repressed wetlands. For Frederic Jameson 'today the past is dead, transformed into a packet of wellworn and thumbed glossy images' in the nostalgia industry. As for the future, he goes on to suggest, 'it is for us either irrelevant or unthinkable' (Jameson 1984, p. 245). Yet for Ballard in *The Drowned World* the future is relevant and thinkable, but only as a return to the past. This return is not nostalgic in the sense of homesickness, but is *unheimlich* or unhomely which is both horrifying and fascinating. What Jameson calls the 'Wagnerian and Spenglerian world dissolutions of J. G. Ballard' should stand as what he goes on to describe as 'exemplary illustrations of the ways in which the imagination of a dying class – in this case the cancelled future of a vanished colonial and imperial destiny – seeks to intoxicate itself with images of death' (ibid.). Or perhaps more precisely in the case of *The Drowned World* with images of parodies of life, of new life being born from the wetland as womb.

These images include the destruction, and rebirth, of the world by water. But not just any water, or even water per se, and certainly not the individualist death by water of, say, Eliot's tired, modernist and mythopoeically nostalgic wasteland. It is the death of western civilisation by wetlands, by the return of its endo- and exo-colonial repressed. This death is also a parodic repetition of its birth from the uncanny. The future is thought, not as difference, but as parody of sameness. The future is thinkable, but only as repetition of or return to the past, both of the species and the individual. There are two kinds of countervailing trajectories at work and in tension here: one entailing a return to the repressed via the uncanny uterus and the other involving the future as a repetition of the uterine past. Perhaps the richness and suggestiveness of Ballard's novel can be attributed in part to the way in which the two lines are used to represent each other.

Part of its fascination can perhaps also be attributed to its quantum interplay with time and space, and the way they are

used to trope each other with colonised places figured as palaeozoic past; post-colonial here and now as pre-colonised and
palaeozoic then and there. Ballard's work is for Jameson 'so rich
and corrupt' because it

testifies powerfully to the contradictions of a properly representational
attempt to grasp the future directly. I would argue, however, that the most
characteristic science fiction does not simply attempt to imagine the 'real'
future of our social system. Rather, its multiple mock futures serve the quite
different function of transforming our own present into the determinate past of
something yet to come. (ibid.)

In particular, *The Drowned World* transforms our own ostensibly
post-colonial and post-modern present into the determinate and
repressed past of the pre-colonial and uterine yet to come. It
enacts the postmodern future anterior, as Lyotard puts it, in which
the present is transformed back into the determinate past of the
pre-modern wetland wilderness yet to come as postmodern wilderness.

In the jacket notes to the Everyman paperback edition of *The
Drowned World*, Kingsley Amis noted that 'this tale of strange and
terrible adventure in a world of steaming jungles has an oppressive power reminiscent of Conrad'. But Ballard's affinity with
Conrad is more than mere reminiscence, more than surface effect
of the iconography of tropical colonialism. Both share a pre-
occupation with the dank heart of darkness of colonialism and the
colonies; they also share a fascination with the western iconography of the lagoon. Ballard's novel is indebted to Conrad's 'The
Lagoon', though it is not infused with quite the same sense of
horror, at least in the later part of *The Drowned World*. In his story
Conrad's anonymous white man travels through black water up a
narrow creek which is described as being

like a ditch: tortuous, fabulously deep; filled with gloom under the thin strip of
pure and shining blue of the heaven. Immense trees soared up, invisible
behind the festooned draperies of creepers. Here and there, near the glistening
blackness of the water, a twisted root of some tall tree showed amongst the
tracery of small ferns, black and dull, writhing and motionless like an arrested
snake. The short words of the paddlers reverberated loudly between the thick
and sombre walls of vegetation. Darkness oozed out from between the trees,

through the tangled maze of the creepers, from behind the great fantastic and unstirring leaves; the darkness mysterious and invincible; darkness scented and poisonous of impenetrable forests. The men poled in the shoaling water. The creek broadened, opening out into a wide sweep of a stagnant lagoon. The forests receded from the marshy bank . . . (Conrad (1898) 1977, p. 172)

The canal through the swamp, the tunnel beneath its over arching canopy, has often been used as a trope in patriarchal western culture, as David Miller puts it, for 'the passageway to the womb' (Miller 1989, p. 25). A journey into the canal or tunnel of the swamp figures for patriarchy the horror of repeating birth in reverse. On these horrific journeys phylogenesis and ontogenesis are repeated in reverse by the patriarchal hero. Phylogenesis in reverse, or going back into the primeval slime of the swamp, repeats ontogenesis in reverse, going back into the black waters of the womb.

Yet Janet Frame tells a different story in which the lagoon represents a way for the first person narrator to establish and celebrate matrilineal descent, particularly a connection back to one of her grandmothers and great-grandmothers. The narrator is thereby gendered as female. The lagoon is not a particularly pleasant place, it is not romanticised, but nor is it psycho-geopathologised, as it is for Conrad, as a place of darkness and horror alone:

at low tide the water is sucked back into the harbour and there is no lagoon, only a stretch of dirty grey sand shaded with dark pools of sea water where you may find a baby octopus if you are lucky, or the spotted orange old house of a crab or the drowned wreckage of a child's toy boat. There is a bridge over the lagoon where you may look down into the little pools and see your image tangled up with sea water and rushes and little bits of cloud. And sometimes at night there is an underwater moon, dim and secret. All this my grandmother told me, my Picton grandmother . . . (Frame (1951) 1991, p. 1)

Although the tide sucks the water from the lagoon back into the harbour, this is not an oral sadistic sucking which enacts a fear of being eaten by (the) (earth) mother, but part of the rhythm of the tides, and of life. At low tide the lagoon is not a place of new life but of death and destruction – the baby octopus who will not survive without water, the shell from a dead crab, the damaged and lost toy left by the kids who play on the edge, who do not venture in like the narrator, the other kids who are anglicised 'Christopher Robins with sand between the toes, sailing toy warships and paddling with bare feet in the pools' (ibid. p. 6), the clean and proper children. Ultimately, as it transpires, the lagoon is the place where the narrator's maori princess and murderess great-grandmother drowned her husband (ibid. p. 6). Yet none of these aspects of the lagoon is dwelt upon, nor psychogeopatho-logised, but are a part of the rhythms of life and death. Frame is part of the minority tradition which sees the wetland as a place of both death and life, darkness and light, as living black water.

The lagoon is not a place which is feared because the self might be absorbed. Rather it is a place where self and other mingle. It is not the pool of Narcissus where the still reflective surface of the water gives back the image of the self as other so that the self can fall in love with itself as other. Nor is it the artificialised wet land of landscape painting which enables the bourgeois viewer to reflect on the surface of things by reflecting on the still surface of the water. Rather 'the water was brown and shiny' (ibid. p. 6), the surface of the lagoon broken up by wind and wave which enables the first person narrator to see both surface (the self) and depth (the other) above and below without falling in love with the self on the surface, nor fearing absorption by the womb of life, nor repressing the depths below, nor desiring sublimation into the heights above. She is situated comfortably and mid-way between surface, depth and height.

The narrator's grandmother instructs her to 'see the lagoon', to see it for what it is, not as a psychogeopathologised projection of fears, even though it is a place of fear, even more so than the bush (ibid. p. 4). But that is part of its fascination – 'the dirty lagoon, full of drifting wood and seaweed and crab's claws. It is dirty and sandy and smelly in summer' (ibid. p. 2). Despite its unpleasant-ness in terms of the conventions of European aesthetics, the narrator 'liked my grandmother to talk about the lagoon . . . But the lagoon never had a proper story, or if it had a proper story, my grandmother never told me' (ibid. p. 2). The lagoon cannot speak for itself, it is always spoken on behalf of for better or worse (usually for worse, but invariably situated at one of two extreme poles), as the wetland is more generally in patriarchal western culture. Frame refuses this either/or logic and deconstructs its binarism to adopt a both/and logic in which the lagoon is both fascinating and unpleasant, rather than fascinating and horrifying. The wetland has never had, and does not want, a 'proper story'; the wetland is the improper.

NOTES

1. A point made by Sofoulis (1991).
2. Paul Carter (1987, p. 84) has argued that 'psychoanalysis borrows its imagery from the realm of space. And who is to say that the spatiality of discourse of the unconscious is merely a figure of speech?' Equally, who is to say that the unconscious as swamp is merely a figure of speech?
3. See Franz Kafka's story 'In the Penal Colony' or variously translated as 'In the Penal Settlement' for a graphic and horrific depiction of the convict's body as literally a surface of inscription for colonial law.
4. I acknowledge here a lecture given by Bob Hodge in a course I taught with him and Zoë Sofoulis entitled 'Language, Culture and the Unconscious' at Murdoch University for alerting me to the cultural possibilities of Freud's theory of symptom.
5. Zoë Sofoulis, personal communication.
6. From 1890–1913 the number of cities in Europe with over 100,000 inhabitants increased by more than a third. See Romein 1978, p. 192. More recently, 'in 1940, one person in 100 lived in a city of one million or more inhabitants; by 1980, one in ten lived in such a city. By the turn of the century almost half of humanity will live in urban centres'. See The World Commission on Environment and Development 1987, p. 16.
7. See McComb and Lake 1990, pp. 13–23. McComb and Lake vary between 'use', 'exploit' and 'harvest' in order to describe Aboriginal interactions with wetlands apparently oblivious to the different connotations which attach to the various words.
8. I am very grateful to Anne Brewster for alerting me to the pertinence of this novel.
9. See The Shorter Oxford English Dictionary, p. 1167.
10. For a graphic illustration see the photographs by Joseph Muench in Niering 1966, pp. 164 and 165, and by Wayne Lawler in McComb and Lake 1990, p. 225.
11. See also Irigaray 1977, p. 70 and 1985, pp. 111 and 224.
12. Similarly, though essentialistically and ahistorically, for Camille Paglia (1991, p. 91) 'female experience is submerged in the world of fluids'.
13. Neumann (1955, p. 48) suggests something along similar lines.
14. Irigaray in several places has propounded a far more positive valuation of the association between lips and female sexuality. See, for example, Irigaray 1977, p. 65.

15. Ruth Barcan, in a paper entitled 'Natural Histories: Nature, History and Gender on the Gold Coast' presented at the Australian Cultural Studies Association Conference in December 1991, argued that 'the Gold Coast, built on the swamp at the edge of the sea, has never fully rid itself of the terror of being re-absorbed'. I am grateful to Karl Neuenfeldt for passing this paper on to me. For Irigaray further on the fluid, see also *Speculum*, p. 237.
16. I am grateful to Cheryl Gole for drawing my attention to Lacan's account.

III

Bogs and Bodies

$\equiv$ 5 $\equiv$

Swamp Sickness: Marsh Miasma and Bodily Effluvia

Wetlands have been associated with disease and death in patriarchal western culture. A derelict bog was described in 1603 as having 'an atmosphere pregnant with pestilence and death' (Coles 1989, p. 153). Over 200 years later Joseph Glynn referred to the Fens as 'the swamp or marsh, exhaling malaria, disease and death' (Darby 1983, p. 176).[1] The idea of wetlands exhaling pestilence or giving birth to death was no mere metaphor in patriarchal western medicine. From at least the fourth or fifth century BCE up to the late nineteenth century, wetlands were thought to breathe out and reproduce malaria, disease and death. Wetlands were a Bachelor Machine for the Bachelor Birth of death.[2] Malaria means literally 'bad air' which was thought to rise from damp or inundated ground (Purseglove 1989, p. 27). The 'miasmatic theory' of disease contended with the contagion theory for two millennia until the parasite carried by anopheles mosquitoes was 'discovered' scientifically in the 1890s and so the 'cause' or vector of malaria established.

In this chapter, I trace and try to deconstruct and decolonise the medical history of the association between wetlands and disease from ancient and classical Greek times to modern times, from the fourth or fifth century BCE to the nineteenth century, from the pioneer of western medicine to the leader of the Sanitary Movement, from Hippocrates to Edwin Chadwick and the mediation of this association in such novels as Dickens' *Martin Chuzzlewit*, Rider Haggard's *She* and C. S. Forester's *The African Queen*, the former two published before the discoveries of the 1890s, the latter after it.

MALARIA, MIASMA AND THE SWAMP OF THE BORA

Ambivalence about wetlands rotates, as it were, around their in-betweenness. Wetlands are neither exclusively land nor purely water, but between land and water, or both land and water, or half land and half water like the Swamp of the Bora the setting for the central chapters of C. S. Forester's *The African Queen* (Forester (1935) 1956, chs 9–12). First published in 1935, so fully forty years after the discovery of the vector of malaria, this 'classic first world war love story' and 'the all-time favourite' according to the cover blurb of the Penguin paperback edition of the novel, perpetuates the miasmatic theory.

In the novel the Swamp of the Bora is described as 'a pestilential fever swamp . . . steaming in a tropical heat, overgrown with dense vegetation, the home of very little animal life, and pestilent with insects' (ibid. pp. 105–6), though insects are animals. It may not have been the home of much charismatic macrofaunal life, but no doubt it was the home of a myriad, but banal microfaunal life. Part of 'the horror of the place' (ibid. p. 137) consists in it being neither land nor water but between the two, as the horror of the slimy for Sartre resides in part in it being between solid and liquid, neither one nor the other. The swamp is between land and water as slime is between solid and liquid (as the sublime is between solid and gas) (see Chapter 2, Figure 1).

The mixing of earth and water marks a transition from the dreariness which in Chapter 2 we saw associated with it to outright horror, especially when this mixture invades the 'African Queen' 'herself':

it was a nightmare time of filth and sludge and stench. Be as careful as they would, the all-pervading mud spread by degrees over everything in and upon the boat, upon themselves, everywhere, and with it came its sickening stench. It was a place of twilight, where everything had to be looked at twice to make sure what it was, so that, as every step might disturb a snake whose bite would be death, their flounderings in the mud were of necessity cautious. Worse than anything else it was a place of malaria. (ibid. p. 138)

Whilst Forester never explicitly states that the stench of the swamp 'caused' malaria, calling it 'a place of malaria' is not equivalent to suggesting it was a place of the mosquito that is the vector for the parasite that infects humans with malaria. Similarly although the swamps of Africa in Rider Haggard's *She* may look bad and be depressing, their worst feature is their smell which were thought to carry disease: 'undoubtedly the worst feature of the swamp was the awful smell of rotting vegetation that hung about it, which was at times positively overpowering, and the malarious exhalations that accompanied it, which we were of course obliged to breathe' (Haggard (1887) 1991, p. 60). The association reproduced here between stench and disease was the foundation of the Sanitary Movement in nineteenth-century Britain. The reference to 'a dank smell of rotting vegetation' in *The African Queen* (Forester (1935) 1956, p. 129) could equally be found in the literature of the Sanitary Movement one hundred years before with its references to 'putrescent animal and vegetable matter'.

To the 'in-betweens' associated with wetlands (in between land and water, air and water, light and dark) we can add miasma as it combines both gas and liquid. Miasma = air (or gas) + water.[3] Miasma, as David Miller has pointed out, are 'an infectious air said to emanate from decaying matter and swamps and marshes' and 'originally stood for any pollution or polluting agent' (Miller 1989, pp. 190, 191–2).[4] Malaria and miasma are literally bad or polluting air which rise from black waters, an appellation, or variations on it, which recurs in *The African Queen* (Forester (1935) 1956, pp. 109, 110, 116, 118, 136, 137) (see Chapter 2, Figure 1). Malaria and marsh miasmata were, in fact, interchangeable terms (Ferguson 1823, p. 273).[5] In 1830 Dr Edwin D. Faust claimed that 'malaria may be considered as an un-combined gas, as the vapour of a volatile solid or liquid, as animalcular, or as a gas combined with water' (Jarcho 1970, p. 36). Malaria, like miasma, = air (or gas) + water.

The name of Faust seems appropriate here, not only for Dr Edwin D., but also for a whole series of doctors and engineers who sold their souls to the Devil of industrial capitalism in exchange for power over wetlands just like Goethe's Faust whose dying speech recounts the Faustian project of discipline and drain in similar terms:

A marshland flanks the mountain-side,
Infecting all that we have gained;
Our gain would reach its greatest pride
If all this noisome bog were drained.
I work that millions may possess this space,
If not secure, a free and active race.
Here man and beast, in green and fertile fields,
Will know the joys that new-won region yields,
Will settle on the firm slopes of a hill
Raised by a bold and zealous people's skill.
A paradise our closed-in land provides,
Though to its margin rage the blustering tides;
When they eat through, in fierce devouring flood,
All swiftly join to make the damage good.
Ay, in this thought I pledge my faith unswerving,
Here wisdom speaks its final word and true,
None is of freedom or life deserving
Unless he daily conquers it anew . . .

                              (Goethe 1969, p. 269)

The freedom and life of enclosed lands have to be won anew each day by conquering the oral sadistic flood which threatens to break the bulwarks erected against it. Paradise is enclosed land, recalling that Eden is a garden. The enclosed lands are located on the slopes of the hill (see Chapter 2, Figure 1). They are thereby placed above and right away from the horrifying and threatening hell of the slimy wetlands with their bad airs and miasma rising from their black waters. The enclosed paradise is also positioned beneath, but not directly below, the terrifying heights of the sublime mountains and sublimated gases.

As heat is the agent which transforms solid into gas (in sublimation) and solid into liquid (in slime), so it is also that which changes liquid into gas (in miasma or malaria in the literal sense of 'bad air'). In Rider Haggard's *She* 'the mist was being sucked up by the sun' (p. 61). In the tropical Swamp of the Bora of *African Queen* 'the heat was colossal' (Forester (1935) 1956, p. 106). Put this together with 'the stifling air' (ibid. p. 106) and the result is 'the stifling heat' (p. 107) but not just any heat and certainly not dry heat. What results, in turn, is 'the dampness of the heat' (p. 107), the heat acting on water to produce 'that inferno of sticky heat' (p. 108) of 'the fever of the day' (p. 116) of the tropical swamp.

For some proponents of the miasmatic theory in temperate climes, such as Adam Seybert writing in the eighteenth century, 'marshes have no noxious influence during the winter season' (Seybert 1799, p. 423).[6] It is only the heat of summer (or of the tropics) which causes the air of marshes to rise and to produce noxiousness. In the temperate zone it is only the blackness of night which produces noxiousness. Gene Stratton-Porter in her novels for adolescents (incidentally with their 'swamp angel' who 'saw in the swamp the garden of the Lord', a paradise to counter Faust's) set in the Limberlost Swamp of northeastern Indiana has one of her characters state that 'the swamp is none too healthy at any time, and at night it is rank poison' (Stratton-Porter (1904) 1990, p. 81, 85, 181). The swamp could, by a stretch of the imagination, be paradise by day, but hell by night, doubly black water, black waters of the night. Night was the time when malignant vapours rose from the swamp.

The prevailing miasmatic view of heat producing pestilence and poison seemed to apply to the tropical zone whereas in temperate climes the view was taken that the blackness of night exacerbates the blackness, and unhealthiness, of the waters. Yet for Seybert's contemporary William Currie it was only the heat of summer which predisposed the human body to malaria by relaxing its solids.[7] As part of the horror of the slimy for Sartre was that it was associated with the relaxing of solids, so for Currie was it part of the horror of malaria. Both entail a slipping away from solidity and from a sense of property and propriety about one's own body, albeit a masculinised one, to something slippery, slimy, evanescent, insubstantial and feminised.

BAD AIRS, BLACK WATERS, UNHEALTHY PLACES

Probably the earliest extant discussion of the association between wetlands and disease in western culture is in the Hippocratic writings. In the essay 'Airs, Waters, Places', the writer is concerned to distinguish 'the effect of different kinds of water, to indicate which are healthy and which unhealthy, and what effects, both good and bad, they may be expected to produce'

(Lloyd 1978, p. 152). Within the category of unhealthy waters are:

stagnant waters from marshes and lakes [which] will necessarily be warm, thick and of an unpleasant smell in summer. Because such water is still and fed by rains, it is evaporated by the hot sun. Thus it is coloured, harmful and productive of biliousness. In winter it will be cold, icy and muddied by melting snow and ice. This makes it productive of phlegm and hoarseness. (ibid. p. 152)

The writer then goes on to enumerate the features of the physiology and physiognomy of swamp dwellers.

The Loeb Classical Library translation of this essay also elaborates the description of the waters in the following miasmatic terms:

turbid, stagnant waters in marshes and swamps are hot, thick and evil-smelling in summer because of their stagnation and failure to flow . . . Waters that issue level on the face of the earth and such as are marshy, standing and stagnant must in summer be hot, thick and stinking, because there is no outflow. (Mattock and Lyons 1969, p. 48)

Not only does the turbidity of swamp waters mean that they mediate between solidity and liquidity (and so violate both these categories), but their stagnancy and hotness violates the qualities of fluidity and coolness generally assigned to waters.

Waters, for the Hippocratic writer, should flow and be cool; waters which do not flow and are not cool violate the order of things. Hot, stagnant and turbid waters are also evil-smelling, whereas cool, flowing and clear waters are pleasant-smelling. Across a range of categories and qualities in the Hippocratic taxonomy of waters, wetlands are ruled out of court. On top of all this, and because of all this, they are bad for bodily health. For what reasons wetlands are bad for health the writer does not say other than that they are 'productive of phlegm' which still begs the question of how stagnant waters produce this condition or substance. There is a certain obviousness that this degree of category- and quality-mediation and violation is such a disruption to the order of things that it must be detrimental to health. Similarly for Pliny the Elder (1991, p. 274) in the first century CE 'doctors investigate which kind of waters are beneficial. They rightly condemn stagnant, sluggish waters, and regard running water as more beneficial because it is rendered fine and healthy by the agitation of the current . . . Wholesome waters should also be without taste or smell.'

ROTTEN AIRS, PUTRID WATERS, UNHEALTHY FENS

This pejorative attitude to wetlands, especially in relation to health, was exported to, or imported into, the United Kingdom and was taken up by its writers in relation to the Fens of England. In Shakespeare's *King Lear*, Lear's final repudiation of Cordelia is couched in the Hippocratic and miasmatic terms of 'Infect her beauty,/ You fen-sucked fogs'. And in *The Tempest*, Caliban curses Prospero in similar terms: 'All the infections that the sun sucks up/ From bogs, fens, flats, on Prosper fall, and make him/ By inchmeal a disease.' The Fens are employed in both instances as a source of insult against the health and beauty or life of someone who has been repudiated or reviled.

The Fens are equated with disease, and its ravages, because of its bad airs. In John Carey's *New and Correct English Atlas* of 1787 the entry for Cambridgeshire remarked that 'in the Isle [of Ely in the Fens] the air is damp, foul, and unwholesome', and for Norfolk that 'the marshy and watery places are aguish and unwholesome' (Darby 1956, p. 176, n. 2). William Camden in his *Britannia* also refers to 'the unwholesome aire of the Fennes' (ibid. p. 24). Yet it is not only the airs which are bad in the Fens. The Fens in fact mix the elements of earth, air and water: 'The moory soil, the wat'ry atmosphere, With damp, unhealthy moisture chills the air. Thick, stinking fogs, and noxious vapours fall, Agues and coughs are epidemical' (ibid. p. 176).

This mixture produces bad health according to William Dugdale: 'what expectation of health can there be to the bodies of men, where there is no element good? The air being for the most part cloudy, gross and full of rotten hairs [*sic*]; the water putrid and muddy, yea full of loathsome vermin, the earth spongy and boggy' (ibid. p. 58). As for the Hippocratic writer, wetlands violate the normative qualities of the elements: air should be clear and dry; water, clear and flowing; earth, dry and solid. Why, as with the previous writers, these violations, these out of court mixings, should lead to unhealth is not made clear. Only a certain obviousness prevails that as they are such a disruption to the order of things across a number of categories, they must bode ill literally.

There were no shortage of writers who bemoaned the unhealthiness of the Fens in accordance with the prevailing miasmatic theory and instances could be enumerated almost *ad infinitum*. If most writers were agreed about the cause of the unhealthiness, they were also agreed about the cure – drain the Fens. The miasmatic theory of disease produced an anti-ecology which destroyed habitats and wetlandscape in a bid ostensibly to improve health. Drainage of the Fens began in the Middle Ages and got going in earnest in the seventeenth century, especially precipitated by the 'Act for the Draining the Great Level Fens'. Fen drainage was to be carried out 'to the great advantage and strengthening of the nation' (ibid. p. 68). The colonising and nationalistic aims of drainage were left in no doubt.

By the nineteenth century, despite the persistence of some fears, health improved. The British historical geographer, H.C. Darby, who published three books on the Fens, remarked that:

despite statistics of longevity, many people [in the nineteenth century] were still 'fearful of entering the fens of Cambridgeshire, lest the Marsh Miasma should shorten their lives.' There was considerable discussion as to what actually caused the ague–humid atmosphere, decaying vegetable matter, or bad drinking water. But whatever the cause might have been, it was certain that the complaint was disappearing. This improvement was generally ascribed to the better drainage. (ibid. pp. 180, 182)

So why enquire further into the cause of the ague when drainage was lessening the incidence? Why not keep draining? It is only in the twentieth century since the vector of the ague has been established that anyone has bothered to enquire into drainage and whether it was really necessary. As Crick concludes in Swift's *Waterland*, 'the problem of the Fens has always been the problem of drainage', but then wonders, 'is it desirable, in the first place, that land should be reclaimed?' (Swift 1984, p. 8).

But why not keep draining when it had such demonstrably positive effects besides the improvement of health? This was certainly the case for John Dyer whose poem 'The Fleece' was first published in 1757:

Yet much may be performed to check the force
Of Nature's rigour: the high heath, by trees
Warm sheltered, may despise the rage of storms:
Moors, bogs, and weeping fens, may learn to smile,
And leave in dykes their soon-forgotten tears.
Labour and art will every aim achieve
Of noble bosoms. Bedford Level, erst
A dreary pathless waste, [but then] . . .
Russel, arose, who drained the rushy fen,
Confined the waves, bid groves and gardens bloom,
And through this new creation led the Ouse,
And gentle Camus, silver-winding streams:
Godlike beneficence: from chaos drear,
To raise the garden and the shady grove
(Darby 1956, p. 162)

Drainage could not only manage a recalcitrant nature and reverse the effects of the Fall; it could also produce a new, unfallen creation and along with returning the earth to the Garden of Eden generate agricultural and moral 'improvements'. Such was the conclusion to which Arthur Young came in 1799 about 'the improvements' following embanking and enclosing in the Fens: 'fens of water, mud, wild fowl, frogs and agues, have been converted to rich pasture and arable'. The result was 'Health improved, morals corrected and the community enriched' (Darby 1983, p. 142). Young enumerates and equates what he sees as the worthless, useless, and the indigenous, if not downright harmful elements of the Fens, against which he poses monoculture and bourgeois culture. It is as if the drainage and discipline of the wetlandscape had a salutary and necessary connection with morality, the former producing the disciplining and draining of the unruly recalcitrance of indigenous populations, the latter restraining evil and inducing good morals.

If stagnant water could be used as a moral emblem as Ruskin did to figure foulness, so could the miasmatic diseases of the wetland be used conversely, as Daniel Webster did in 1851, to trope secession and disunion as:

a region of gloom, and morass, and swamp; no cheerful breezes fan it, no spirit of health visits it; it is all malaria. It is all fever and ague. Nothing beautiful or useful grows in it; the traveller through it breathes miasma, and treads among all things unwholesome and loathsome. (Miller 1989, p. 10)

If secession and disunion could be seen as swamp, the swamp could conversely be troped as disorder and rebellion as Harriet Beecher Stowe did in 1856 in *Dred; A Tale of the Great Dismal Swamp*:

the reader who consults the map will discover that the whole eastern shore of the southern states [of the United States], with slight interruptions, is belted by an immense chain of swamps, regions of helpless disorder, where the abundant growth and vegetation of nature, sucking up its forces from the humid soil, seems to rejoice in a savage exuberance, and bid defiance to all human efforts to penetrate and subdue. (Stowe 1856, p. 189)

Northern economic and political supremacy could be legitimated by southern swampiness. Similarly, proto-Fascist violence was necessary for Ernst Jünger because of the miasmatic corruption of the Wiemar Republic whose atmosphere he describes as 'like a swamp which can only be purified by explosions' (Hermand 1980, p. 125).

This connection between drainage and moral correction became the whole premise of the Sanitary Movement of the following century. It is certainly a conjunction which has persisted for a long time, even right into the standard history of technology which describes drainage in terms of 'training and regulating', almost like the toilet training of some infantile and recalcitrant child (Harris 1957, p. 300). The history admits, though, that 'there was no imperative need to reclaim the land for purely agricultural reasons' and puts it down 'to the often despised profit-motive that we owe' the drainage, and enclosure, of the Fens (ibid. pp. 315–16). Like the Swan River settlement away from home, the Fens were 'a colony for capitalists' at home, especially in the light of the class robbery of the enclosure movement.

Yet all writing about the Fens did not exclusively praise drainage. Probably the most passionate, articulate and ambivalent writer about the Fens was Charles Kingsley. One of his *Prose Idylls* is devoted to the Fens and provides a highly ambivalent account of their drainage in which the discourses of the rationalist and the naturalist, the pastoral and the pastoralist contend against each other:

> reason and conscience tell us, that it is right and good that the Great Fen should have become, instead of a waste and howling wilderness, a garden of the Lord . . . And yet the fancy may linger, without blame, over the shining mere, the golden reed-beds, the countless water-fowl, the strange and gaudy insects, the wild nature, the mystery, the majesty . . . which haunted the deep fens for many a hundred years. (Kingsley 1884, p. 95)

Unlike his predecessor Arthur Young who was happy to see everything and anything indigenous to the Fens transformed by drainage into monoculture, Kingsley both praises the benefits of so-called progress brought by drainage and bemoans simultaneously the losses inflicted on the Fens. Kingsley seems to want his cake of the old, pre-drainage Fens, and to eat the cake of the new, drained Fens, too. Or perhaps more precisely, he seems to want the good of the old Fens, such as the native flora and fauna, and not the bad, such as harshness and sickness, and to want the good of the new Fens, and not the bad.

But as the good of the new Fens is tantamount to the loss of the good of the old Fens, Kingsley is stuck in a bind, though for him the advantages of drainage ultimately outweigh the disadvantages: 'ah, well, at least we shall have wheat and mutton instead [of the native flora and fauna], and no more typhus and ague; and, it is to be hoped, no more brandy-drinking and opium-eating; and children will live and not die. For it was a hard place to live in, the old Fen' (ibid. p. 97).

In the Fens, opium was known as 'the antidote to the effect of the noxious vapours' (Dobson 1980, p. 370). Yet despite improvements in the health of children, prosaic wheat and common mutton, even the eradication through drainage of what Kingsley calls 'the cold and poisonous fogs' (Kingsley 1884, p. 102) rising from the Fens, seem for him to be a poor swap, an inequitable exchange, for what he also calls poetically 'the shining meres, the golden reed-beds, the countless water-fowl, the strange and gaudy insects, the wild nature, the mystery, the majesty' (ibid. p. 95).[8] Similar sentiments were expressed at about the same time by Gerard Manley Hopkins:

What would the world be, once bereft
Of wet and of wildness? Let them be left,
O let them be left, wildness and wet;
Long live the weeds and the wilderness yet.
                         (Hopkins 1953, p. 151)

This was the minority view in England, though it is one that was also propounded at about the same time in the United States as I will show in the concluding chapter.

For Kingsley, however, unlike Hopkins, the bourgeois virtues of cleanliness, healthiness, godliness and temperance, outweigh his mourning and ambivalent nostalgia for the lost wet and wild Fens: 'it is all gone; and it was good and right that it should go when it had done its work, and that the civilisation of the fen should be taken up and carried out . . . till "out of slough and bogs accursed, he made a garden of pleasure"' (Kingsley 1884, p. 123). The fen is 'a waste and howling wilderness' to be transformed into a pastoral and pastoralist Eden, a barbarous place to be civilised by drainage, but as Thoreau said 'a howling wilderness does not howl: it is the imagination of the traveller that does the howling' (Thoreau (1864) 1988, p. 300). 'Waste' and 'howling' are qualities produced by fertile imaginations and ascribed to 'wilderness', not qualities of the land itself. They are cultural constructions of nature.

Kingsley argues that the Fens constituted something of a first in unifying populations against nature as a common threat and a test-case of 'man [with large-scale industrial technology on 'his' side] versus ['the brute powers' (Kingsley 1884, p. 129) of] nature':

it was in the fens, perhaps, that the necessity of combined effort for fighting the brute powers of nature first awakened public spirit and associate labour, and the sense of a common interest between men of different countries and races. But the progress was very slow; and the first civilisers of the fen were men who had nothing less in their minds than to conquer nature. (ibid. pp. 117–18)

The drainage of the fens also gave for Darby the phrase 'man [sic] and his conquest of nature' its 'full meaning' (Darby 1956. p. 28).

The conquest of wetlands by drainage was a militaristic invasion and imperialist colonisation which were no mere metaphors as Swift's *Waterland* shows by its references to the 'endless and stationary war against water' (Swift 1984, p. 289), and to waging 'war on water, mud' (ibid. p. 60). Wetlands are a soft target, an easily vanquished foe. The war against wetlands was also no mere metaphor for Benito Mussolini in the 1930s with his scheme for draining the Pontine Marshes. These marshes have been described as recently as the 1950s in a typical reductive gesture as 'nothing but a malaria-infested marsh' (Harris 1957, p. 311). They were subject to various attempts to drain them from the fourth century BCE. Goethe had warned about their 'dangerous miasma', though he thought they 'do not actually look as dreary as people in Rome usually describe them' (Goethe 1970, pp. 180–1). For Mussolini though, the beneficial and 'peaceful' task of reclaiming the 'never-ending fen' of the Pontine Marshes was 'the kind of war which we prefer' (Pini 1939, p. 180), perhaps because it was so easy to win with the tools, or weapons, of modern technology. According to Rachelle Mussolini, Benito's wife, to win 'the battle against the swamps', especially the Pontine Marshes, was one of his five main objectives (Mussolini 1959, p. 88). Benito achieved this objective and later boasted that his two main achievements were that he had made the Italian trains run on time and that he had drained the Pontine Marshes.[9]

Italian fascist drainage of the Pontine Marshes merely extended and heightened processes already at work in modernity and modernisation at large, in particular through the industrial conquest of the earth. German proto-fascism, on the other hand, internalised and psychologised these processes implicit in modernity, in particular via the sublimation of the slimy as we saw in Chapter 2. Rather than the bad dream of the right and the whipping-boy of the left, Fascism is the boys' own wet daydream of modernity and modernisation which became a nightmare. Fascism was what Theweleit calls a 'filiarchy' (Theweleit 1987, p. 108), the rule of the sons, and part of 'the regime of the brother' of modernity as Juliet Flower MacCannell calls it (MacCannell 1991). This rule operates in the boys' network, not so much the old boys – the patriarchy, the fathers – but the adult and middle-aged boys who have usurped the power of the fathers and rule over their younger brothers and sisters. Some of that power was enacted in and around marshes since they posed a threat in Fascist political aesthetics to the masculinist clean and proper body. The Fascist 'condestruction' of marshes was modernity and modernisation writ large.

The chief weapon used in the war against wetlands was drainage which has not been seen as destructive and regressive but as 'a major manifestation of progress' (Purseglove 1989, p. 28). For Crick in Swift's *Waterland* 'my humble model for progress is the reclamation of land' (Swift 1984, p. 291) with its 'curious password: Drainage' (ibid. p. 65). Whether 'progress' is a good or bad thing is another matter. Perhaps it was a 'good thing' then but a 'bad thing' now. In the recent novel *Soundings* by Liam Davison, a kind of Australian *Waterland*, drainage 'put men in control' (control of what? one wonders, though the intransitivity might have been all that was desired) because it was 'Science put to a practical use' which produced 'a vision of Progress. It showed how the land could be mastered' (Davison 1993, pp. 97, 99, 103, 172). The effects of drainage on the (wet)landscape were profound. In the Netherlands alone between 1590 and 1640 some 70,000 acres were drained (Schama 1988, p. 38). Clarence Glacken argues that 'one of the truly great landscape changes in modern times has been marsh, bog, marine and lacustrine drainage' (Glacken 1967, p. 348). I would go even further and suggest that wetlands' drainage is the landscape change which quintessentially characterises modernity and modernisation with all its cultural, corporeal and psychological conditions of possibility (or 'causes'), its ramifications and its widespread effects.

THE GREAT DISMAL, THE BLACK AND OTHER AMERICAN SWAMPS

When Europeans, raised with and habituated to an experience of the flat openness of the Fens or of the Lowlands – the Netherlands – confronted the swamps, the wooded wetlands, of the New World and of other colonies they were in for a cultural shock of the most profound magnitude, one with durable after-effects. One of the earliest surviving accounts of such contact is from the eighteenth century describing William Byrd's encounter with the Great Dismal Swamp. Like Joseph Glynn writing about the Fens as 'exhaling malaria, disease and death' (Darby 1983, p. 176), Byrd describes how 'the foul damps [of the Great Dismal Swamp] ascend without ceasing, corrupt the air, and render it unfit for respiration' (Byrd 1966, p. 194). These, in turn, give rise, to what he calls 'the agues and other distempers occasioned by the noxious vapours that rise perpetually from that vast extent of mire and nastiness' (ibid. p. 70). These agues and other distempers could even lead to death or not quite with the inhabitants not only possessing 'a cadaverous complexion', but the living looking like the walking dead, like zombies:

the exhalations that continually rise from this vast body of mire and nastiness infect the air for many miles round and render it very unwholesome for the bordering inhabitants. It makes them liable to agues, pleurisies, and many other distempers that kill abundance of people and make the rest look no better than ghosts. (ibid. p. 202)

Like the Fens, the Great Dismal Swamp mixed the elements of earth, air and water to such an extent that for Byrd 'the soil was so full of water and the air so full of damps that nothing but a Dutchman could live in them' (ibid. pp. 222–3). By mixing the elements the wetland becomes 'this dirty place' with 'the badness of the water and . . . the malignity of the air' (ibid. p. 191) made up of 'the noisome exhalations that ascend from that filthy place' (p. 199). Wetlands are not only dirty but also dirt in the sense of Mary Douglas' famous definition as 'matter out of place' (Douglas 1966, p. 35). The wetland is matter out of place par excellence. Wetlands have been seen as solid in the place of liquid, water in the place of air, heat in the place of coolness, dark in the place of light, night in the place of day and death in the place of life.

The wetland is matter out of place in all the major western categories: solid, gas, heat/light and liquid. Wetlands mix all four western elements of matter: earth, air, fire and water. Even with their living waters wetlands have been constituted as dead matter. The typical response to the dirtiness of wetlands has been to try and put matter back in its proper place wherever possible and where it is impossible to denigrate their deviation from the norm. Drainage, for instance, is putting the earth and water of the wetland back in their respective, and respectable, places. In the process wetlands are put in their subjugated place and drylands substituted in their place instead.

The solution for Byrd to the problem of the Great Dismal Swamp, like his compatriots with the Fens, was drainage though he recognised that 'it would require a great sum of money to drain it, but the public treasure could not be better bestowed than to preserve the lives of his Majesty's liege people and at the same time render so great a tract of swamp very profitable' (Byrd p. 202). Like Kingsley, Byrd is driven by a capitalist, monocultural imperative. But unlike Kingsley, Byrd is not ambivalent about the Great Dismal Swamp. George Washington, however, was. He became involved in a project to drain the Swamp, yet saw it as both 'a glorious paradise, abounding in wildfowl and game' albeit 'subject to wets and unhealthiness' (Davis 1962, p. 48; Washington 1976, p. 323).

No such ambivalence was felt by some writers about other American wetlands such as the Black Swamp in north-west Ohio. For one writer, this swamp is a 'forsaken, desolate, ague-smitten, tangled and inhospitable wilderness whose diseases spread with rapidity and relentless mortality' (McGarvey 1988a, p. 65). Wilderness did not evoke then the warm inner glow it produces today (Nash 1982; Oelschlaeger 1991). Similar sentiments about the Black Swamp were voiced more recently: 'an abysmal, human-forsaken swamp ... infested with mosquitoes, gnats, horseflies, crawling with snakes, moccasins, water rats and prowled by ravaging wolves and wildcats, no wonder the first white people who saw the place called it the Great Black Swamp' (McGarvey 1988b, p. 99). Although not quite a God-forsaken place, the Black Swamp was definitely seen in detrimental terms by white humans as a locus of death, disease and evil. Like the Great Dismal Swamp, the very name of the Black Swamp evokes the pejorative.

The Americans seem to have an ignoble tradition of inventing or giving pejorative names to wetlands with overtones of despair or despondency such as the Great Dismal Swamp, or associations of racism such as the Black Swamp, or just plain expressions of doggone tiredness as in Dorothy Langley's novel *Swamp Angel* set in 'the Missouri swamp settlement' of Weary Water which is described as 'a sinkhole of infection, malaria, typhoid and dysentery' (Langley 1982, p. 15). Yet the most extensive literary meditation on the association between disease and American wetlands is undertaken by Dickens in Chapter 33 of *Martin Chuzzlewit* set in the swamp at the confluence of the Mississippi and Ohio Rivers where 'the night air ain't quite wholesome' but is, as in Stratton-Porter's *Freckles*, 'deadly poison'.

The confluence of the two rivers is also the site for the settlement of Cairo which was evoked ironically by Herman Melville in *The Confidence Man* in the following miasmatic terms:

at Cairo, the old established firm of Fever and Ague is still settling up its unfinished business; that Creole grave-digger, Yellow Jack – his hand at the mattock and spade has not lost its cunning; while Don Saturnus Typhus, taking his constitutional with Death, Calvin Edison and three undertakers, in the morass, snuffs up the mephitic breeze with zest. (Melville 1990, p. 156)[10]

What Melville was to ironise, and even send up in a proto-postmodernist, sardonic sneer, his contemporary, Walt Whitman, was to repudiate in prosaic terms: 'the winds are really not infectious' (Miller 1989, p. 210).

Dickens transforms Cairo in *Martin Chuzzlewit* into the heavily ironised and 'thriving city of Eden [which] was also a terrestrial Paradise, upon the showing of its proprietors'. This Eden is a kind of nightmarish, hyperbolic hypothetical of a pre-drained paradise which, in the hands of Dickens, becomes the locus for a scathing diagnosis of the ills of nineteenth-century American boosterism. Dickens is highly critical of the rhetoric of real estate advertising and the discrepancy between it and the reality to which it ostensibly refers and supposedly represents. One of the most unhealthy aspects of Eden is what Dickens calls with typical irony 'the salubrious air of Eden', like the health-preserving airs of Mudfog. This physical air not only produces the metaphysical 'air of great despondency and little hope on everything', but also gives Martin a good dose of malaria in accordance with the then prevailing miasmatic theory.

Yet the experience for Martin is a rite of passage, a kind of post-epic descent into the underworld – a secular version of Dante's skirting the 'sullen souls stuck in slime' of the fifth circle of Hell – in his journey toward middle-class manhood:

in the hideous solitude of that most hideous place, with Hope so far removed, Ambition quenched, and Death beside him rattling at the very door, reflection came, as in a plague-beleaguered town; and so he felt and knew the failing of his life, and saw distinctly what an ugly spot it was. Eden was a hard school to learn so hard a lesson in; but there were teachers in the swamp and thicket, and the pestilential air, who had a searching method of their own. (Dickens (1844) 1953, p. 506)

Eden is a school of hard knocks, a university of life, though this does not diminish the fact that 'rain, heat, foul slime and noxious vapour, with all the ills and filthy things they bred, prevailed. The earth, the air, the vegetation, and the water that they drank, all teemed with deadly properties'.

STAGNANT POOLS AND THE SANITARY MOVEMENT

These fears about the association between stagnant waters and malarial disease were one of the bases of the nineteenth-century Sanitary Movement which began in Great Britain under the tutelage of Edwin Chadwick and quickly spread to the colonies. (Bolton and Hunt 1978, pp. 1–17).[11] According to Peter Stallybrass and Allon White, Chadwick's *Report on the Sanitary Condition of the Labouring Population of Great Britain* first published in 1842 was 'an instant best-seller, and more than 10,000 copies were distributed free' (Stallybrass and White 1986, p. 125). As Chadwick was a former literary assistant to Jeremy Bentham (Illich 1986, p. 45)[12], he seems to have wanted to practice the principles of utilitarianism on the city. The Sanitary Movement applied Bentham's Panopticon (literally 'to see everything') principle, initially a design for a circular prison with a central observation tower, to the hitherto invisible underside of the city, whether it be working class areas or to sewerage, in order to render it, or at least its effects, visible to health inspectors. By reading the effects in accordance with the interpretive grids of the Sanitary Movement, the causes could be deduced and remedies prescribed. The Movement was also based on the utopian and educative impulse that good drainage meant good morals. It subscribed to the 'discipline and drain' school of thought. It was thus part of the general move in the nineteenth century to police, regulate and place under surveillance the life of populations (Foucault 1979).

The founding trope of the Sanitary Movement was the working class abode as tropical swamp. Southwood Smith, described as 'perhaps the most influential of the medical pioneers' of the Sanitary Movement, drew the analogy in his *Treatise on Fever* published in 1830:

the room of a fever-patient, in a small and heated apartment in London, with no perflation of fresh air, is perfectly analogous to a stagnant pool in Ethiopia, full of the bodies of dead locusts. The poison generated in both cases is the same; the difference is merely in the degree of its potency. Nature with her burning sun, her stilled and pent-up wind, her stagnant and teeming marsh, manufactures plague on a large and fearful scale. (Greenwood 1953, p. 504)

Just as for Adam Seybert in the previous century, heat is the agent which produces malignancy. And so it is in this century in, for example, *The African Queen* in which the sun blazes malignantly (Forester (1935) 1956, p. 111). The other within – the inner-city slum, the working class – is analogous to, or the same as for all intents and purposes, as the other without – the tropical swamp, 'the natives'.

As 'nature' and 'the natives' were colonised by European imperialism with its medical and other technologies, including engineering, so were the tropical swamp and the inner-city slum. One other within, the working class, was even seen by Chadwick as, and reduced to, agricultural animals, a yet further colonised other within: 'the precautions applicable to the sheep and cattle will be deemed equally applicable to the labouring population' (Chadwick (1842) 1965, p. 158). Although Engels did not draw the same analogy as Southwood Smith, he used the same metaphor in describing the East End of London as 'an ever-spreading pool of stagnant misery and desolation' (Engels (1882) 1969, pp. 31, 32, 34). By a neat, but not nice, historical irony the area of the city built on drained marshes reverts to the swamp of working-class urban areas. Such is the persistence of the association between disease and the working-class 'swamp' that when Anthony Wohl recently introduced Southwood Smith's analogy he did so by describing 'the slum dwelling as potentially lethal as any disease-ridden tropical marsh' (Wohl 1984, p. 286). Similarly Eric Hobsbawm has luridly referred to what he calls 'the slimy pools of sweated and casual poverty' (Hobsbawm 1985, pp. 249, 232).

The Sanitary Movement was clearly based on the miasmatic theory of disease: for it, as for Forester later in *The African Queen*, and as M. W. Flinn put it in his introduction to the recent edition of Chadwick's book, 'smells generated disease' (Chadwick (1842) 1965, p. 62). Indeed, for Chadwick, 'all smell is disease' (Stally-brass and White 1986, p. 139).[13] This conclusion must both have been the result of and have entailed a massive repression and sublimation of the olfactory in which all smells were seen as bad, though generally it was only bad smells which were thought in the miasmatic theory to have caused disease. For the Sanitary Movement disease was thought, as Flinn also put it, to have been 'generated in the miasma given off by decaying organic matter' (Chadwick (1842) 1965, p. 62).

The archetypal and typical place where this generation occurred was, of course, the wetland: 'miasmata are admitted by the concurring testimonies of medical practitioners in every part of the globe, to be produced by the action of the sun upon low, swampy ground' (ibid. p. 155). Miasma could also arise from human effluvia; indeed, for all intents and purposes, they were one and the same thing. Miasma and effluvia went together, whether it be the miasma from 'putrescent animal and vegetable matter' (compare 'a dank smell of rotting vegetation' in *The African Queen* (Forester (1935) 1956, p. 129)) or the effluvial 'excretions from the human body, accumulated and corrupting' (Chadwick (1842) 1965, p. 63). The representation of the human body, or at least the nether regions of it, as swamp is a typical trope to which I return in the following chapter. Miasmata and effluvia were interchangeable terms, and effluvia were described invariably as noxious, as were vapours (Currie 1799, pp. 129, 131, 135, 139).[14] Indeed, as we have already seen in a number of instances, 'noxious' is the usual adjective used to modify 'vapours' in the miasmatic theory. It was the nineteenth-century equivalent of our sense of 'toxic'.

Effluvia arose not only from the human body but also from wetlands. Indeed, Thomas Wright in 1794 could even refer to 'the effluvia of swampy lands ... producing ague as an epidemic' (Wright 1799, p. 243). For Sir Arthur Faulkner in 1820 'malaria ... is only another name for marsh effluvia' (Jarcho 1970, p. 35). As colonial marshes and swamps have been colonised by surveying and engineering, draining or filling, so bodily effluvia have also been colonised by medicine. Certain parts of our bodies and certain bodily effluvia are, Klaus Theweleit argues, 'colonial territories, colonised phenomenon' (Theweleit 1987, p. 416). Or more precisely, the concept of effluvia is a colonising device which equates the body with swamps to be colonised, its fluids to be disciplined, drained or canalised and finds bodily odours repulsive so requiring that they be masked beneath and by deodorants or perfumes.[15] Decolonisation of nature not only entails the decolonisation of the bush and the swamp, but also the decolonisation of the body through a kind of corporeal demapping in which the body would relocate itself in relation to the nether regions, both its own and those of the earth.

Such was the dominance of the miasmatic theory that Edward Harrison could even claim universality for it in the early nineteenth century through the spread of European imperialism and its medicine. Not only malaria but also typhus as Melville's *Confidence Man* shows was thought to be caused by miasma. Harrison found that 'the connection between humidity and the rot [or typhus (p. 158)] is universally admitted by experienced graziers' (Chadwick (1842) 1965, p. 154). Medicine and agriculture concurred happily, and universally, in ascribing the cause of typhus to the miasma rising from some 'unenclosed fen'. Harrison's view can be seen manifested shortly after in Chapter 9 of *Jane Eyre* by Charlotte Brontë first published in 1847: 'that forest dell, where Lowood lay, was the cradle of fog and fog-bred pestilence; which, quickening with the quickening spring, crept into the Orphan Asylum, breathed typhus through its crowded schoolroom and dormitory, and . . . transformed the seminary into a hospital'. Lowood was thus transformed into 'the seat of contagion' with its 'effluvia of mortality'.

The cure for Chadwick was simple, drainage along with garbage collection (Chadwick (1842) 1965, p. 99). Or, if wetlands were the cause of the disease, then in the terms of one of Chadwick's authorities, it was a simple matter of 'suppress the marsh to expel the fever,' the sloganeering cure prescribed by the Sanitary Movement (ibid. p. 161). Drainage produced the double benefit of salubrious air and productive soil. It could even be applied successfully to the age-old problem of the Isle of Ely. Chadwick quotes approvingly John Marshall:

it has been shown that the Isle of Ely was at one period in a desolate state, being frequently inundated by the upland waters, and destitute of the adequate means of drainage; the lower parts became a wilderness of stagnant pools, the exhalations from which loaded the air with pestiferous vapours and fogs; now . . . , by the labour, industry, and spirit of the inhabitants, a forlorn waste has been converted into pleasant and fertile pastures, and they themselves have been rewarded by bounteous harvests. Drainage, embankments, engines, and enclosures have given stability to the soil . . . as well as salubrity to the air. These very considerable improvements, though carried on at a great expense, have at last turned to a double account, both in reclaiming much ground and improving the rest, and in contributing to the healthiness of the inhabitants. Works of modern refinement have given a totally different face and character to this once neglected spot. (ibid. pp. 150–1, see also p. 153)

In other words, modernisation has converted primeval wilderness and modern wasteland into pastoral productivity. Unlike Kingsley who bemoaned the loss of the old Fens, Marshall is completely gung-ho about drainage.

As for Arthur Young in the eighteenth century, so for the Sanitary Movement in the nineteenth: suppressing the marsh and draining the fen not only expelled fevers, improved health and produced fertile land for agriculture, but also reformed morals. Chadwick asked 'how much of rebellion, of moral depravity and of crime has its root in physical disorder and depravity' (Stallybrass and White 1986, p. 131). He went on to postulate that 'the fever nests and seats of physical depravity are also the seats of moral depravity, disorder and crime with which the police have most to do' (ibid. p. 131). Unlike Southwood Smith who only drew a parallel between the spaces of the tropical swamp and working-class abode, Chadwick goes further to draw an analogy between the physical and moral.

For Chadwick, not only were 'the powerful operation of depraved domestic habits [seen] as a predisposing cause to disease' (Chadwick (1842) 1965, p. 205), but also 'the noxious physical agencies depress the health and bodily condition of the population and act as obstacles to education and to moral culture' (ibid. p. 268). The moral could be corrected by correcting the physical. The way to a 'man's' heart' was not only through his stomach, but also through his nose. In order to foster moral culture, it was simply a matter of draining stagnant pools: 'the cases of moral improvement of a population, by cleansing, draining and the improvement of the internal and external conditions of the dwelling' (ibid. p. 200). In this way 'the effects of the noxious physical agencies on the moral condition of the population' (ibid. p. 204) could be ameliorated.

By a poignant historical irony, or perhaps poetic justice, the engineering which destroyed wetlands through drainage, and so had some initial ameliorating effect on the incidence of malaria, has contributed also to the re-introduction of malaria through the building of dams and the creation of artificial lakes, ideal breeding grounds for mosquitoes, a conclusion which at least one writer in the nineteenth-century had foreseen (Collins 1829, pp. 30, 31)[16] and about which one recent writer has warned (Waddy 1975, p. 8). The position of engineering in regard to wetlands encapsulates the contradictory nature of modernisation and modernity: on the one hand, the power of engineering to alter the (wet)landscape, to drain wetlands and so to ameliorate malaria; and on the other the power to build dams, to create artificial lakes and so to re-introduce malaria. This paradox also highlights a shift in the social status of the engineer from modern hero to postmodern Faust who has sold his (and usually it is his) soul to the Devil of industrial capitalism in return for power and money at the expense of, amongst other places, the wetlands, their indigenous populations of people, fauna and flora, along with the mystery and majesty of wet wilderness.

NOTES

1. See also Darby 1974, p. 9.
2. For the history of malaria see Covell 1967, pp. 281–5; Bruce-Chwatt 1976, pp. 168–76; 1988, pp. 1–59; and Harrison 1978. I am grateful to Jim Warren for alerting me to the fact that the University of Western Australia Medical Library has a good collection of books and articles on malaria and miasma.
3. See Ladurie 1989, p. 1. This book is concerned with the bourgeois capitalist conquest of water as a generalised principle abstracted from particular waters such as wetlands. I would argue that this abstraction is complicit with the conquest of waters.

4. See also Parker 1983, especially pp. 1–17.

5. See also Curtin 1964, pp. 182–8. I am grateful to Jim Warren for drawing my attention to Curtin, who discusses Ferguson.

6. Despite his adherence to the prevailing miasmatic theory of the time, Seybert was a remarkable proto-wetlands conservationist whose positive concluding comments on marshes (p. 429) will be discussed in the concluding chapter.

7. For William Currie (1799, p. 137) 'the intermittent fever seldom attacked any but whose solids had been previously relaxed by the preceding heat of the summer'.

8. See also Kingsley 1877, pp. 12–13.

9. Toby Miller, personal communication. I am grateful to him for pointing out the pertinence of Mussolini to my work.

10. See also Miller 1989, pp. 185, 202, 203; and Sussman 1978, pp. 32–56.

11. I am grateful to Toby Miller for drawing my attention to this article.

12. This book is similar to Goubert's (see above n. 3) in that it is largely concerned with the bourgeois capitalist conquest of water as a generalised principle abstracted from particular waters such as wetlands and is thus arguably complicit with it.

13. As Roy Porter points out, 'stench was, in fact, disease'. See his 'Foreword' to Corbin 1994, p. vi. Corbin discusses miasma, effluvia, marshes and swamps extensively but, surprisingly, makes no reference to malaria.

14. See also Allen 1947, p. 492.

15. See Chapter 6 below for an extensive discussion of the relationship between the body and swamps.

16. I am grateful to Ned Rossiter for drawing my attention to this book.

6

The Nether Regions: Sexuality, the Gendered Body and the Swamps of the Great Mother/ Earth

Wetlands have been, and still are to some extent, associated in patriarchal western culture with disease, and with its possible, fatal outcome. They can even be associated not so much with the end of life as with lifelessness and the uninhabitable. The swamp has not only been regarded as a 'region of death' as David Miller puts it, but also as 'a convincing valley of the shadow of death', as Miller also argues, in which no 'living thing' could live, or would want to live, and where to do so would be inviting death (Miller 1989, pp. 23, 25). From the swamp of the shadow of death it is only one small step to the hell of the shades of the dead which have also been associated with wetlands in patriarchal western culture. The mire, the bog, the swamp is the environmental *memento mori*, the reminder of death. 'A festering corpse beneath the mire' (Soyinka 1973, p. 83) is the environmental analogue to the skull beneath the skin.

In this chapter I trace and try to deconstruct and decolonise the association between the wetland and the lower or 'nether' regions of death, hell and 'the body', a body which is always gendered (as well as differentiated along lines of class and ethnicity). Mikhail Bakhtin called the 'nether regions' of 'the body' 'the grotesque lower bodily stratum' (Bakhtin 1984). Equally there is a grotesque lower earthly stratum, the nether wetlands. There is a strong association in patriarchal and filiarchal western culture between the nether regions of the body and the nether lands of the physical and metaphysical wetlandscape.

This association between the swamp, hell and the lower body is usually of an excremental, misogynist and hom(m)osexual

character (paradoxically both homophobic and homo-erotic).[1]
Although these associations are often reproduced uncritically, I
argue that they can be employed in parodically carnivalesque
inversions which nevertheless leave in place the hierarchical
privileging of the terms inverted. They can also, however, be used
in politically resistive subversions, or at least can be read as such,
which subverts the hierarchy of normative proprieties itself. Such
subversion takes place in T. Coraghessan Boyle's coruscating short
story 'Greasy Lake'. Equally such associations can be symptomatic
of a repression of the maternal and anal, and their displacement
into fantasies of denial and disavowal of the maternal, as in the
Puritan Milton's revisionist *Paradise Lost*.

This is not to deny that the swamp, or the wetland more
generally, is a place of death – it is. But it is also a place of life.
Importantly, wetlands are places of both life and death. They are
living black waters where, as David Miller puts it, 'the intermingling of life and death' takes place (Miller 1989, p. 80). Or as
Marguerite Duras' narrator puts it in *The Sea Wall [Un barrage
contre le Pacifique]* set in the Mekong Delta, there is 'a chaotic,
creation-of-the-world intermingling' in the Delta. And so it is in
the wetland more generally (Duras 1986, p. 126).[2] Perhaps what
makes the swamp so horrific is the fact that, as Miller goes on to
suggest, 'the swamp is an ambiguous image that evokes both
begetting and destroying' (Miller 1989, p. 238). Death is necessary
for life to be reborn. In the midst of death and decay in wetlands,
we are in the midst of new life being reborn.

Modern western culture has largely dismissed the association of
wetlands with life, as living waters, and concentrated exclusively
on the association of wetlands with death, as black waters. Yet the
patriarchal western association of wetlands with death, disease,
excrement and hell is belied by recent ecological understandings
of wetlands which identify them as areas of vital, life-giving
significance. Although they may be degraded through human
interference and mismanagement, in their indigenous condition
they are not only 'among the most productive ecosystems on
earth', but also arguably among the most important ecosystems for
sustaining human and other life by purifying water (Mitsch and
Gosselink 1986, p. 11).

THE SWAMP OF THE SHADOW OF DEATH

For William Byrd in the eighteenth century the Great Dismal Swamp was not only a place of disease, but also 'a miserable morass where nothing can inhabit', including humans: 'those nether lands were full of bogs, of marshes and swamps, not fit for human creatures to live in' (Byrd 1966, p. 174). Nor were they fit for any other animal creature to live in for that matter. In a taxonomic gesture Byrd dispenses with the entire animal and plant kingdoms: 'no living creature could inhabit that inhospitable place' (ibid. p. 68).

For Byrd not even a creature from the Netherlands could survive in the 'nether land' of the Great Dismal Swamp:

since the Surveyors had entered the Dismal, they had laid eyes on no living creature: neither bird nor beast, insect nor reptile came in view. Doubtless the eternal shade that broods over this mighty bog and hinders the sunbeams from blessing the ground makes it an uncomfortable habitation for anything that has life. Not so much as a Zeeland frog could endure so aguish a situation. (ibid. pp. 194, 197)

The unproblematic slippage from 'the eternal shade' beneath the overarching canopy of the swamp to the eternal shades, the ghosts of the dead, was easily and repeatedly made.

This view of the Great Dismal Swamp as a region of death, or at least lifelessness where life was construed in narrow terms by being reduced to and equated with animal life, persisted into the following century when Frances Anne Kemble wrote in her journal that: 'it looked like some blasted region lying under an enchanter's ban, such as one reads of in old stories. Nothing lived or moved throughout the loathsome solitude, and the sunbeams themselves seemed to sicken and grow pale as they glided like ghosts through these watery woods' (Miller 1989, p. 281, n. 27). Construed in these terms the swamp is the antithesis of the national park forest. It is a kind of anti-cathedral where a black mass is conducted.[3]

For Kemble, as for Byrd, the Great Dismal Swamp was an uninhabitable place, a region of death and even a wilderness:

into this wilderness it seems impossible that the hand of human industry, or the foot of human wayfaring should ever penetrate, no wholesome growth can take root in its slimy depths; a wild jungle chokes up parts of it with a reedy, rattling covert for venomous reptiles; the rest is a succession of black ponds, sweltering under black cypress boughs – a place forbid. (Miller 1989, p. 281)

The swamp here is a forbidden place of black water steaming beneath the black sun of depression and melancholia.

Hubert Davis begins his book on the Great Dismal Swamp by stating it was written 'for those who think of the Dismal Swamp as a hostile filthy mire, teeming with poisonous snakes, and cloaked in mystery and misunderstanding' (Davis 1962). Against this view Davis maintains that 'Dismal Swamp is a well-balanced small world of strange, mystic charm, teeming with reptiles, large and small animals, beautiful birds, delicate aromatic wildflowers, giant trees, and beautiful ferns' (ibid. p. 13). Wetlands have been seen recently as either a place of teeming life or a region of creeping death, either an uninhabitable and impenetrable swamp or a life-giving and enjoyable wilderness. The challenge today is to see wetlands as regions of both life and death, as living black waters, in a kind of postmodern double vision which is both poetic (but not romanticist) and ecological (but not mechanistic).

Yet even attempts to uphold the ecological values of wetlands often lapse back into standard swampspeak. Davis does his own bit to contribute to the 'mystery and misunderstanding' in which the Swamp is 'cloaked' by claiming that 'Dismal Swamp, much like an alluring and seductive woman, is capricious, contra-dictory, with tantalising beauty and oft changing moods. She invites familiarity, yet metes out swift and cruel punishment to those who are lured into her web of dangers'.[4] For Davis the Swamp is both *femme fatale* who lures men, and it usually is men, to their death in her seductive embrace, and spider woman who entraps them in her dangerous webs of intrigue. Not only is 'she' 'cloaked in mystery and misunderstanding', but also 'in womanly fashion she dons a different ensemble with jewels and fragrances for every mood' (ibid. p. 14).

Yet beneath the enticing and decorous surface of her cloaks and jewels lie her 'undergarments' covering her 'nether regions'. These underclothes are exposed by frost, rains and occasional hurric-anes which 'pluck one by one her various garments from her

fading glory to expose an under-raiment of somber brown, bright-
ened by patches of evergreens' (ibid. p. 15). The dark and dirty
underside of the swamp is repressed temporarily beneath its
glamorous and glittering surface only later to resurface in the
figure of the aging, down-at-heel spider woman beneath whose
superficial charms there still lies death and decay.

The strongest association between a swamp, a *femme fatale* and
a spider woman was made in Rider Haggard's late nineteenth-
century, best-selling adventure-romance *She* with its eponymous
phallic woman, 'a white sorceress living in the heart of an African
swamp'. She is also a spider woman who weaves 'the web of her
fatal fascination' into which male characters are drawn, against
which they must heroically fight and from which they manfully
emerge with their manhood vindicated (Haggard (1887) 1991,
pp. 46, 177). Africa is feminised, or more precisely femme fatal-
ised, and made to wear 'a belt of swamp all along the East African
coast' (ibid. p. 49; see also p. 124). The male hero has to negotiate
his way through this belt of black waters, has to undo it and throw
open the 'cloak of mystery' beneath it in order to allow the light of
reason to penetrate the 'dark continent' of Africa and of female
sexuality. This dark continent is essentially a land of black waters,
inevitable disease, possible death and certainly the uncanny (ibid.
pp. 17, 77, 142, 238). In order to achieve the object of his quest,
the hero has to overcome his fear of 'the poisonous exhalations
from the marsh' (ibid. p. 67) which arise from it and infect in
accordance with the prevailing miasmatic theory of disease of the
day.

Because of the threat of disease from intercourse (sexual and
verbal) with 'her', the dark continent is almost inevitably a land of
the dead, or worse 'a land of swamps and evil things and dead old
shadows of the dead' (ibid. p. 143). Not only is the swamp here 'a
region of death', not only 'a valley of the shadow of death' as in
the Psalmist's words, not only a land of the shadows of the dead,
it is also a land of the dead-old shadows of the dead and so triply
horrifying. The swamp has been viewed as a necropolis in
western culture. In *She* the swamp is viewed as city of the dead
ghosts of the dead.

Death in the swamp was a nineteenth-century romantic idea of
union with mother nature. Following Marie Bonaparte, Gaston

Bachelard remarks how 'nature is for the grown man, an immensely enlarged, eternal mother, projected into infinity'. Emotionally, nature is a projection of the mother'. (Bachelard 1983, p. 115; Bonaparte 1949, p. 286). More specifically, the wetland emotionally is a projection of the mother's womb often entailing an intimate, yet horrifying, encounter as a result of which the mother is rejected. The experience of the swamp, David Miller argues, 'in at least a few instances in Mid-Victorian America . . ., involved a more intense encounter with the overwhelming body of "mother nature" ' (Miller 1989, p. 104). By contemplating death through plunging into a bog, the romantic was ostensibly seeking union with infinity and the void, but he was also fleeing from what he saw as the stifling, (s)mothering embrace of (the) mother (nature) and enacting his desire to dissolve his own ego and body boundaries.

The swamp was also romanticised in nineteenth-century American culture as a mad flight from the living mother to the dead maiden as, for example, in Thomas Moore's ballad 'The Lake of the Dismal Swamp' (Moore, pp. 242–3). Writing of this poem David Miller has argued that 'the maiden in her ethereal femininity personifies the lake that is her sanctuary; she is its *genius loci*' (Miller 1989, p.43). If this is the case, then the romantic man who seeks sanctuary in the lake or swamp is seeking it in or with the maiden in her sanctuary. Unlike the nymphs who were, in the words of Mircea Eliade, the Greek divinities of 'all flowing waters, of all springs and of all fountains', and unlike 'the (living) lady of the (deep) lake' who was the British Romantics' figure of 'la belle dame sans merci,' the dead maiden of the shallow lake or swamp is the local goddess of all stagnant waters, of all dismal swamps, the figure of '*la jeune fille (mort) sans merci*' for American Romantics (Eliade 1958, p. 204). All are deeply ambivalent figures; associated with life and death, love and madness, they are *femmes fatale* and *vitale*. In discussing the nymph, Mircea Eliade describes 'the ambivalent feeling of fear and attraction to water which at once destroys (for the 'fascination' of the nymphs brings madness, the destruction of the personality) and germinates, which at once kills and assists birth' (ibid. p. 205). The besotted young man who wanders into the dismal swamp in search of his lost young maiden living or dead, or in flight from her or his mother, has gone mad having lost himself, or contemplating losing himself, in the dismal swamp in search for, or flight from her.[5]

The fear of being buried in a bog resurfaces (excuse the pun) in the twentieth century from its very first year. Sir William MacGregor, governor of Nigeria and medical man, wrote in 1900 that 'I have an invincible, selfish, irrational objection to being buried in these swamps' of Lagos (Harrison 1978, p. 130). Presumably he wanted to be buried 'at home' in the dry land. Yet one of the fears that moves William Byrd's surveyors was 'a dread of laying their bones in a bog that would soon spew them up again' (Byrd 1966, p. 197). It was not so much a dread of being buried in the bog that horrifies them as the fear that the bog will not accept them – ingest, digest and excrete them – but will reject them, regurgitate them.

The fear of being buried in mud, or wet land, beset many American soldiers serving in Vietnam during the 1960s. A veteran recalled that 'I wanted to die clean. It didn't matter if I died – but I just didn't want to die with mud on my boots, all filthy. Death wasn't so bad if you were clean.' Death was horrific if you were dirty, even worse if you were wet and dirty, in a word, muddy. The ultimate and recurring nightmare of one Viet-vet was that 'I would end up shot, lying along the side of the road, dying in the mud.' Psychotherapist Robert Jay Lifton comments of the fear of 'dying in the mud' that it 'meant dying in filth or evil, without reason or purpose – without nobility or dignity of any kind' (Lifton, p. 222).[6]

Perhaps the ultimate horror would be to die in stinking wet dirt, in a word, excrement. Another Viet-vet related how:

I heard of one helicopter pilot in Nam who was carrying a shit-house on his helicopter. He crashed and was killed, and was buried under the whole shit-house and all the shit. I thought that if I was going to die in Vietnam, that's the way I would like to die. I didn't want to die a heroic death. That was the way to die in Vietnam. (Lifton, pp. 222–3)

This would be the ultimate anti-heroic death, to die buried in shit. Lifton comments that 'only excrement, ultimate filth, could provide the appropriate burial ground in Vietnam – the appropriate symbolism for dying' (ibid. pp. 222–3). A filthy death was the only appropriately 'filthy truth' about a filthy war. A muddy death, a death in a wet land or in excrement, on the one hand, is feared as the ultimate horror by the heroic who desire a clean and proper death, and on the other hand is desired as the appropriate gesture by the anti-heroic who desire a filthy and improper death as the ulitmate act of rebellion, of snubbing anti-septic convention, and as a way of representing the truth about a shitty war.

The anti-heroic and anti-social possibilities of the wetland associated with death and decay have recently been explored and exploited by T. Coraghessan Boyle in his short story 'Greasy Lake' with its cast of 'greasy characters' (Boyle, 1986). Set in a time 'when it was good to be bad' (ibid. p. 1) and on the outskirts of a middle-American town, the story takes place more precisely at, and sometimes in, the eponymous lake of the title. The first-person narrator and central character relates how the Indians called Greasy Lake Wakan:

a reference to the clarity of its waters. Now it was fetid and murky, the mud banks glittering with broken glass and strewn with beer cans and the charred remains of bonfires . . . We went up to the lake because everyone went there, because we wanted to snuff the rich scent of possibility on the breeze, watch a girl take off her clothes and plunge into the festering murk, drink beer, smoke pot, howl at the stars, savour the incongruous full-throated roar of rock and roll against the primeval susurrus of frogs and crickets. This was nature. (ibid. p. 2)

Rather than nature, though, this was nature heavily loaded with irony, a modern wetwasteland worked over by corporate capitalism into dead black water which is made to serve anti-social ends, even to resist the cultural construction of nature as benign and pristine. The wetland fulfils this anti-social function of site of youthful resistance as it is the outside and the underside of good citizenship and the city or town.

In 'Greasy Lake' the narrator and his two buddies become 'dissociated from humanity and civilisation' (ibid. p. 6) by descending into the barbaric, rapine hell of 'lust and greed and the purest primal badness' (ibid. p. 6) of the underside of middle, moral majority America. When a fight and an attempted gang rape perpetrated by the narrator and his buddies is interrupted by newcomers, the narrator flees through 'the feculent undergrowth at the lake's edge' into the 'muck and tepid water' (ibid. p. 6). He wades into the lake until:

the water lapped at my waist as I looked out over the moon-burnished ripples, the mats of algae that clung to the surface like scabs ... I waded deeper, stealthy, hunted, the ooze sucking at my sneakers. As I was about to take the plunge ... I blundered into something. Something unspeakable, obscene, something soft, wet, moss-grown ... I understood what it was that bobbed there so inadmissibly in the dark. Understood and stumbled back in horror and revulsion ... but the muck took hold of my feet ... and suddenly I was pitching face forward into the buoyant black mass, throwing out my hands in desperation while simultaneously conjuring the image of reeking frogs and muskrats revolving in slicks of their own deliquescing juices. AAAAArrrgh! I shot from the water like a torpedo, the dead man rotating to expose a mossy beard and eyes cold as the moon. (ibid. pp. 6, 7)

The narrator emerges from the water 'coated in mud and slime and worse,' and from 'the hideous stinking embrace of a three-days-dead-corpse' (ibid. p. 7). He lies on the edge of the lake smelling 'the bad breath of decay all around me' and feeling 'the primeval ooze subtly reconstituting itself to accommodate my upper thighs and testicles' (ibid. p. 9) in a grotesque parody of either or both sexual intercourse with Mother Earth and/or an evolutionary birth from primal slime. The primeval slime of Greasy Lake, or any modern wetwasteland for that matter degraded by agriculture, industry or urban development, is not a pure, watery womb of life associated with vitality and nourishment – in short, living waters – but a sullied, muddy tomb of death associated with decay and excrement – dead black waters.

THE BOWELS OF THE EARTH, THE KIDNEYS OF THE ENVIRONMENT OR THE PLACENTA OF THE GREAT MOTHER/EARTH?

The association between wetlands and waste, the aquaterrestrial and the excremental is not that arbitrary or bizarre when one recalls that, in the words of two wetlands scientists:

our waters are ... used for waste disposal. Stormwater comes rapidly off roofs and roads, discharging at high speed into ponds or streams; processed or raw sewage is discharged into some water bodies; industrial effluents containing heavy metals, oil and grease may pass through pipes into our lakes and nearshore ecosystems; heavy metals find their way into wetlands from roadways, in seepage from rubbish dumps; nutrients and other contaminants pass into groundwater and so to our wetlands from agricultural catchments, septic tanks and urban refuse. (McComb and Lake 1990, p. 12)

Culturally, wetlands are often represented as the cloaca of the city and the bowels of the earth to which human waste and waste-products run, or if not that wetlands are at least connected to the bowels of the earth via the underground aquifer of which they often are the expression and on which the city is reliant for domestic water supplies. In 1929 it was maintained that Perth (Western Australia) 'lies in an artesian basin, and in the past [and still in the present] much of the domestic water supply has been drawn from the bowels of the earth' (Colebatch 1929, p. 355).

Yet wetlands ecologically are, in the words of two other wetlands' scientists, 'sometimes described as 'the kidneys of the landscape' for the functions they perform in hydrologic and chemical cycles and as the downstream receivers of wastes from both natural and human sources' (Mitsch and Gosselink 1986, p. 3). In other words, the 'kidneys of the (wet)landscape' are not the end-point of the environmental waste-disposal system, not the bowels, but the 'sewage treatment plant', to use an engineering, mechanical metaphor, of the body environmental. The kidneys of the human body 'excrete urine and so remove effete nitrogenous matter from blood'.[7] The kidneys of the body of the earth perform the same function in relation to water, the life-blood of the earth.

Yet at present the production of human nitrogenous waste is far exceeding the capacity of the environmental kidneys to deal with it. The result is that wetlands are becoming nutrient-enriched or eutrophic. Eutrophication is affecting the whole body environmental with the result that algal-blooms give rise to offensive odours, 'cause' botulism in birds and midge-population explosions which are a nuisance to humans living nearby. Instead of malaria, wetlands should now be associated with the diseases inflicted upon them by human abuse and mismanagement. Eutrophication should be redefined as a disease of wetlands 'carried' or 'caused' or vectored by humans. The body environmental is suffering from kidney failure due to humans disposing of too much nitrogenous material into its excretory system; wetlands are becoming dead black waters.

Yet rather than the kidneys of the environment or the bowels of the earth it may be more apposite biologically and politically to think of, and trope like the later Kerans of Ballard's *Drowned World*, the wetland as the placenta of the Great Mother/Earth. The placenta, for Hélène Rouch, like the wetland,

constitutes a system regulating exchanges between the two organisms [of the mother and the foetus], not merely quantitatively regulating the exchanges (nutritious substances from mother to fetus, waste matter in the other direction), but also modifying the maternal metabolism: transforming, storing, and redistributing maternal substances for both her own and the foetus' benefit. (Rouch in Irigaray 1993, p. 39)

These functions are also performed by wetlands for the benefit of life, including human life, on earth. Wetlands both regulate the exchanges of nutrients and wastes between present life and new life and modify the natural environment. The placenta is what Rouch (1987, pp. 71–9) elsewhere calls the 'third party' which mediates between the mother and the foetus as the wetland mediates between the Great Mother/Earth and humans, between 'old' and new life, between land and water, between river and sea and so on.

If the earth with its alimentary canals, its swampy anuses, its environment kidneys and its placental wetlands can be likened to the body, the body can equally be likened to the earth with its kidney wetlands or sewerage works and wetland placenta. As Crick, the history school-teacher narrator of Graham Swift's novel *Waterland*, reminds us, 'there is such a thing as human drainage too, such a thing as human pumping' (Swift 1984, p. 27) which can be found not least in the brain. In fact, for Crick the narrator there is an analogy between 'the topography of the Fens' and 'the even more intricate topography of the medulla and the cerebellum, which have their own networks of channels and ducts and their own dependence on the constant distribution of fluids' (ibid. pp. 68–9). In other words, the cerebellum and medulla are a fen of the body human.

The brain can even have its swamps. In *The Sexual Brain* Simon LeVay has argued that:

people tend to stay away from the hypothalmus. Most brain scientists (including myself until recently) prefer the sunny expanses of the cerebral

cortex to the dark, claustrophobic regions at the base of the brain. They think
of the hypothalmus – though they would never admit this to you – as haunted
by animal spirits and the ghosts of primal urges. They suspect that it houses,
not the usual shiny hardware of cognition, but some witches' brew of slimy,
pulsating neurons adrift in a broth of mind-altering chemicals. (LeVay 1992,
p. 39)[8]

In other words, the hypothalmus has been figured as a swamp, or
more specifically as the Great Dismal Swamp in terms very similar
to Frances Kemble's description. The hypothalmus is also the site
of LeVay's own controversial discovery of brain differences
between homosexual and heterosexual men.

For the staid and traditional brain scientist the swamp of the
hypothalmus is a horrific, unhomely place to steer clear off. In
short, it is 'a dark continent'. It would be far easier and safer to
stay at home in the dry land of the cerebral cortex with its shiny
and sublimated goodies than to flounder around in the wet land of
the hypothalmus with its dirty and slimy baddies. Yet this
representation of the hypothalmus is necessary for LeVay to
constitute himself as the hero, albeit patriarchal and filiarchal,
who is prepared to penetrate the swamp, the 'dark continent' of
hypothalmic homosexuality, which no 'man' has been brave
enough to tread, or sink into, before.

Rather than deconstruct this metaphorisation, LeVay reinforces
it in cognate terms by inviting the reader to descend 'to this
underworld' like Kerans descending into the drowned underworld of London. It seems as though LeVay has been reading too
much science fiction, such as *The Drowned World*, and not
enough feminist theory. By feminising the hypothalmus as
'womby vaultage,' by reproducing the feminised figure of the
swamp, and by associating both with the underworld, LeVay
seems to be ascribing some sort of essential effeminacy, or at least
patriarchal and filiarchally constructed femininity, to homosexuality, to be portraying women in misogynist terms and to be
representing the swamp in placist terms.

THE SWAMP AS HELL

Despite, or as well as, its attractions for the romantic sensibility in
modern western culture the swamp has been seen as about as

close as one can get to hell on earth. In fact, what Bruce McGarvey calls 'a myth of the hellish swamp' and 'myths of the swamp as Hell' have been around in western culture for a long time (McGarvey 1988a, pp. 59, 61). In our increasingly secularised post-Christian and post-modern age and world, though, it is more the myth of the swamp as hell than Hell as swamp which persists today.

The myth of Hell as swamp goes back at least as far as the ancient Greeks. In his history of Dutch drainage, Johannes van Veen suggests that for the ancient Greeks the Netherlands coast was where 'Hades and the Gates of Hell were supposed to be' (van Veen 1949, p. 12). The Netherlands were literally the ultimate nether lands, even of the human body. In the seventeenth century they were described as 'the buttock of the world' (Schama 1988, p. 44). Van Veen goes on to suggest that 'Homer supposes Hell to be there also,' though more recent attempts to map Odysseus' journey have him going nowhere quite so far afield (Homer 1967, p. 341). Nevertheless, Hades, the land of the dead, is a wetland for Homer:

[at] . . . the limit . . . of the deep-running Ocean
There lie the community and city of Kimmerian people,
hidden in fog and cloud, nor does Helios, the radiant sun, ever break
   through
the dark to illuminate them with his shining,
neither when he climbs up into the starry heaven
nor when he wheels to return again from heaven to earth,
but always a glum night is spread over wretched mortals.
                                                    (Homer 1965, p. 168)

Similarly when Lucretius (1994, p. 13) described the nature of the universe in the first century BCE he referred to 'the murky depths and dreary sloughs of the underworld'.

Later Dante was to rework the River Styx, one of rivers of the Classical underworld, into a slimy swamp. His influence on Milton has often been remarked, but whether this includes their similar views of wetlands and their association of hell with them is another matter for Milton saw the Styx in *Paradise Lost* as 'the burning lake', as 'abhorrèd' and as 'the flood of deadly hate' (II, 576, 577). In the same book he elaborates later in more detail on Hell as a place of

... lakes, fens, bogs, dens and shades of death,
A universe of death, which God by curse
Created evil, for evil only good,
Where all life dies, death lives, and Nature breeds,
Perverse, all monstrous, all prodigious things,
Abominable, inutterable, and worse
Than fables yet have feigned, or fear conceived,
Gorgons and Hydras, and Chimeras dire.

(II, 621–8)

This view of Hell as swamp (and of the slime as unutterable) persists through the intervening centuries well into the twentieth century. It is present in *The African Queen* with its reference to 'that inferno of sticky heat' when describing the Swamp of the Bora (Forester (1935) 1956, p. 108). As the swamp has also served as a figure for the unconscious, it is only one small step to complete the circle of inference by referring to 'the inferno of the unconscious' (Kristeva 1989, p. 206).

In a similar vein 'stygian' gets overworked as an adjective in relation to the darkness and gloom of swamps. Milton used it in *Paradise Lost* to refer to 'the Stygian pool ... that obscure sojourn' (III, 14–15). In 1864 a federal officer made a journey by night into the Great Dismal Swamp and reported 'the stygian darkness of the forest' (Pugh and Williams 1964, p. 10). The adjective is also present in Peter Tremayne's trashy horror story with its typical reference to 'the Stygian gloom of the mangrove swamp' (Tremayne 1985, p. 186). Yet mangrove swamps can also be places of soft light refracted through their golden green leaves.

Most recently the view of wetland as hell can be found in Peter Matthiessen's *Killing Mister Watson*, a kind of American *Waterland*. Set in the Florida Everglades, this factive novel can be read as an attempt to deconstruct the myth of the swamp as hell from within through its play of narratorial voices, yet in the process it reproduces some other myths such as the psychogeopathological (a mind-and-land pathology) association between the anal, the excremental and the aquaterrestrial. One of the narrators, Henry Thompson, wonders of one family of early settlers 'with all of Florida to choose from, who else would come to these overflowed rain-rotted islands with not enough high ground to build an outhouse, and so many skeeters plaguing you in the bad summers you thought you'd took the wrong turn to Hell' (Matthiessen 1990 pp. 10, 11).

When hurricanes wreak havoc across this wetlandscape they do not merely reveal the brown 'under-raiments' beneath the normally placid surface of the fatal margins (hardly a shore). Rather, they overturn the privies and turn up all that is rotting beneath the surface:

in the hurricane's wake, the labyrinthine coast where the Everglades' deltas meet the Gulf of Mexico lies broken, stunned, flattened to mud by the wild tread of God . . . In the dark air a sharp fish stink is infused with corruption of dead animals and blackened vegetables, of excrement in overflowing pits from which shack privies have been blown away. (ibid. pp. 1, 2)

This is hell because God has trodden on it and damned it to destruction; these are, according to another narrator in the novel, 'godforsaken swamps' (ibid. p. 214) though previously the same narrator had claimed that some 'used to call this place God's country, and we still do, cause nobody but God would want no part of it' (ibid. p. 56).

The swamp is hell not only because God has forsaken it and trodden upon it, but also because it is the place of timelessness, the place where time seemingly stands still. In Vereen Bell's 1941 novel *Swamp Water* set in the Okefenokee Swamp of Georgia 'the swamp was a lost world; a nether land where the crawling things in the muck and the screaming things in the air had triumphed, and man rotted in the peaty earth; the space was the space of eternity, endless, changeless' (Bell (1941) 1981, p. 212). The wetlandscape is represented here as the landscape of eternity, the space of timelessness; wetlandscape as endless timescape (see Chapter 2, Figure 1). Ecologically this is nonsense as the wetland is arguably the most dynamic and most mutable of all ecosystems. Rather than being timeless, wetlands are 'timeful'. Wetlands are the space of mutability. David Rains Wallace has pointed out that 'wetlands are intermediaries between past and future as well as between water and land' (Wallace 1987, p. 207). In fact, wetlands constantly change. If it does not change, a wetland dies. A static wetland is a dead wetland.

The wetland is the space of eternity not because it is changeless but because it is ever-changing, ever life-reproducing and life-restoring. The only constant is change, and the wetland is the exemplar of this principle. As Wallace goes on to summarise:

> various studies have indicated not only that salt marshes produce most of our seafood, but that swamps store water with much less evaporation than reservoirs, that marshes and swamps purify sewage effluent more cheaply than treatment plants, and that floodplain forests regulate stream flow more cheaply than flood-control reservoirs. Wetlands are not simply living museums for the preservation of alligators and storks, but a vital part of the hydrologic system that produces clean and reliable water supplies. (ibid. p. 216)

Yet those wetlands which are habitats for alligators and storks, crocodiles and jabirus, alone and do not have any immediately exploitable part to play in the hydrologic system should still be preserved, not just for their limited, utilitarian benefits to humans, but also for their benefits for other species, for their aesthetic, ecotourist, 'passive' or eco-recreational and spiritual benefits to humans, and even just for their own sake.

The wetland is integral to the evolutionary process, and so it defies or upsets the traditional reading of the Judeo–Christian creation story. Referring to the Black Swamp of Ohio one early settler remarked that 'we read that God divided the land from the water, but here is a place He forgot' (McGarvey 1988a, pp. 59, 61). God could be seen to have forgotten or neglected all wetlands. God created everything and so created wetlands. But when He divided the land from the water, He did not divide all land from all water. He left wetlands undivided. Wetlands as both land and water, or as areas between slipped through. Hence, they were not only an aberration but also a vestige of, or a throwback to, the first day of creation when everything was chaotic, disordered, and the spirit of God brooded on the face of the waters.

The wetland harks back to the time when the entire earth was wetland. Genesis 1:1 needs to be rewritten or retranslated in postmodern ecological terms which recognises the importance, even primacy, of wetlands: 'in the beginning God created the heavens and the earth, and the earth was wetlands'. But perhaps when God divided the land from the water it was not that He forgot or forsook wetlands rather that He knew that the land and the water and the creatures that He later created to live in or on them would not survive and would not be able to reproduce without wetlands. Subsequent days of creation do not supersede and forget rather their predecessors but build and rely upon them. In the Judeo–Christian creation story the days have been seen in terms of a progressive, supersessive time-line heading towards the ultimate goal of humanity, and not as an evolutionary process.

Without the first day of creation, the creation of the earth as wetland, there is no sixth day, no creation of humankind. Without the recreation of wetlands on the first day, and every day for that matter, there is no recreation of human life; there is no human life at all, on the sixth or any other day. Rather than being forgotten by God, the wetland is God's first, and for many best, work, without which no other day could ever begin, could ever dawn. Rather than being hell, the wetland is the closest thing to heaven because it was created right after heaven. When one goes to the wetland one goes back to the first day of creation when it springs fresh and new from the hand and word of God, when the spirit broods over the earth as wetland like a hovering marsh or swamp harrier not like a malevolent ghost or monster.

The Judeo–Christian tradition holds a grain of present day ecological knowledge about the role of wetlands if it is read critically, but generally it has been at the forefront of the denigration and destruction of wetlands. In the vanguard of those moves is the perhaps unlikely propagandist, of John Milton, the revisionist Puritan, who wrote in *Paradise Lost* about:

> ... this wild abyss,
> The womb of Nature and perhaps her grave,
> Of neither sea, nor shore, nor air, nor fire,
> But all these in their pregnant causes mixed
> Confus'dly, and which thus must ever fight,
> Unless th' Almighty Maker them ordain
> His dark materials to create more worlds,
> Into this wild abyss the wary Fiend
> Stood on the brink of Hell.
>
> (II, 910–18)

In a later chapter I will argue that Milton's Satan is an archetypal swamp monster. For Milton, like Friedrich Nietzsche as Luce Irigaray has remarked, 'holes mean only the abyss for you' hence the need to invaginate and penetrate surfaces as a way of denying and repressing the depths of the Body of the Mother (Irigaray 1991, p. 7). Traditionally the extension of the swamp has been invaginated so it can be penetrated by phallic heroes. A journey across or through a swamp penetrating its invaginated surface denies or represses the Body of the Mother and its depths. In patriarchy, to use Lyotard's words, 'there are no holes, only invaginations of surfaces' (Lyotard 1993, p. 21)[9] as a denial and repression of the holes and depths of the mother's body.

God, for Milton, brings order out of the wetland chaos, out of 'the vast immeasurable abyss/Outrageous as a sea, dark, wasteful, wild' (VII, 211, 212):

> ... on the wat'ry calm
> His brooding wings the Spirit of God outspread,
> And vital virtue infused, and vital warmth
> Throughout the fluid mass, but downward purged
> The black tartareous cold infernal dregs
> Adverse to life ...
>
> (VII, 234–9)

Hell as swamp was not just any part of hell for Milton, but the lowest part of the underworld. Milton reproduces part of what has been called 'the naive view of the world' in which Tartarus was 'the lowest part of the underworld' (Kirk et al. 1983, p. 9). In *The Iliad* 'misty Tartaros' is 'the deepest gulf below earth' and in Hesiod's *Theogony* it is 'the grievous, dank limits which even the gods detest' (ibid. pp. 9, 40). Milton regards water as lifeless until the spirit (or breath for they are the same word in Hebrew and Greek) of God breathes life into it. He rejects both 'Thales' idea of the earth floating on water' and the idea that water is 'in some way the source of all things' (ibid. pp. 11, 88).

For Milton the Spirit of God replaces water in the pre-Socratic cosmogony. Whereas in the latter, 'water is the continuing, hidden constituent of all things' (ibid. p. 94), for the former, the Spirit of God becomes this secret, life-giving essence. Whereas in the latter, 'all living things depend on water for nourishment' (ibid. p. 94) for the former all living things depend on the Spirit of God for life. Whereas the latter regard 'the nurture of all things to be moist' (ibid. p. 89), for the former the nurture of all things is dry and windy. Milton's position is not only indicative of a sublimation away from the materiality and immanent spirituality of water to the transcendental spirituality of God, but also of a fundamental shift in the conceptualisation of human life from being constituted by fluids to being indicated by breath (and Spirit), and of the human body from being regarded as a system of fluids, or wetland (a view retrieved in Graham Swift's *Waterland*), to a bag of winds, of hot air (see Chapter 2, Figure 1). As Luce Irigaray has remarked in her imaginary dialogue with Nietzsche, 'it is always hot, dry and hard in your world' (Irigaray 1991, p. 13) as it is in Milton's and for the Stoics for whom 'all the elements of the universe are sustained by heat . . . it provides the life-force and is the source of all that comes to be' (Cicero 1972, p. 134). In the wetland world of the mother, it is always cool, wet and soft. The wetland, like the forest for John Evelyn, is 'the moist womb of the world' (Thoreau 1993, p. 66). The wetland is the life-force and source of all that comes to be.

THE GREAT MOTHER

The swamp as Hell, as *femme fatale* and as spider woman, coalesce in the swamp as the mouth of mother earth which can ingest and as the bowel of mother earth which can excrete. Erich Neumann in his archetypal and misogynist study of the Great Mother referred to 'hell as the devouring maw of the earth' (Neumann 1955, p. 171). Swamps have also been regarded as devouring. For example, Louis Simonis described the Black Swamp of Ohio as 'a land which devoured its inhabitants' (McGarvey 1988a, p. 65). In Liam Davison's *Sounding* 'whole mobs of cattle were swallowed up' by the swamp (Davison 1993, p. 45). In Graham Swift's *Waterland*, after Crick puts the tip of his index finger into Mary's hole, he finds it necessary to put it in further and then to put a second finger in, because 'Mary's hole began to reveal a further power to suck, to ingest; a voracity' (Swift 1984, p. 43), like the slime for Sartre. Crick finds Mary's hole orally sadistic, like the Fens themselves which can suck and ingest a human body into their gaping and ravenous maw.

Yet the journey into the gaping and ravenous mouth of the swamp is regarded as heroic in patriarchal and filiarchal cultures.

In Gene Stratton-Porter's novels set in the Limberlost Swamp her narrator refers to 'the green maw of the quagmire' and to 'the black swamp-muck' in which one could be 'swallowed up in that clinging sea of blackness' (Stratton-Porter (1910) 1988, p. 10 and (1904) 1990, p. 22). Typically in mythology, in Erich Neumann's words, 'the underworld, the earth womb . . . is experienced in the archetypal nocturnal sea voyage of the sun or the hero' (Neumann 1955, p. 157). Typically in post-epic narratives, especially in imperialist adventure-romances such as Rider Haggard's *She* and Forester's *The African Queen*, the underworld, the earth womb, is experienced as an archetypal swamp journey of the hero.

If the swamp is the mouth of hell in western culture, it is equally 'the backside of the world' as Milton described it in *Paradise Lost* (III, 494), or as Nat Turner puts it in his fictionalised *Confessions*, 'the hellish bowels of the earth' (Styron 1981, p. 333). It is the place where, as a seventeenth-century doctor imagined it, 'the excrements of the earth, the slime and scum of the water' are gathered and then expelled (Thomas 1984, p. 55). The swamp has been figured as the entire alimentary canal from the beginning to the end, especially the lower or nether end, the place where matter passes between inside and outside; the place that can ingest, digest and excrete, or ingest, repulse and regur-gitate; the place that will accept and transform, draw nourishment and reproduce life, or revolt and expel, gag and spew. In graphic terms, by taking a sagittal section of the human body as a metaphorical map of Vietnam, the swamp is the Mekong delta of the human body running down to its 'mouth' which is also anus in a patriarchal and filiarchal conflation of oral and anal sadism.

Besides going down the alimentary canal, a journey into 'the moral geography' (Schama 1988, ch. 1 esp. p. 24) of the swamp can also go up the birth canal which leads to the womb where life is recreated. Sandra Gilbert and Susan Gubar remark of Haggard's *She* that:

as the men make their way inland, through vaporous marshes and stagnant canals, the [wet]landscape across which they journey seems increasingly like a Freudian female *paysage moralisé* ['moralised landscape' or more precisely moralised wetlandscape or *marécage moralisé*] . . . and, in a symbolic return to the womb, they are carried up ancient birth canals through a rocky defile into 'a vast cup of earth' that is ruled by *She-who-must-be-obeyed*. (Gilbert and Gubar 1989, p. 13)[10]

This cup is 'a spot where the vital forces of the world visibly exist', 'the very womb of the world' (Haggard (1887) 1991, pp. 29 and 279) like the black bowl of the lagoon in Ballard's *Drowned World*.

The destination of the heroic journey up the birth-canal is, in Neumann's, words, 'the womb of the earth'. In patriarchal and filiarchal culture, the source of life is transformed into an orally sadistic hell, into what Neumann calls 'the deadly devouring maw of the underworld . . . the abyss of hell, the dark hole of the depths, the devouring womb of the grave and of death, of darkness without life, of nothingness' in a blatant and disavowing fantasy of reversal and negation (Neumann 1955, p. 149). Womb becomes tomb in what Alice Jardine (1985, p. 32) calls '*the* founding fantasy: the active negation of the Mother' presaging or based on and repeating *the* founding act of what Irigaray (1993, p. 11) calls 'the murder of the woman-mother'. This fantasy essentially involves two inversions: the idea of tomb as womb becomes womb as tomb, and the notion that death is not the end of life as such but the beginning of new life becomes the idea that new life is a kind of death.[11] There is a hell of a lot of difference culturally and politically between the tomb as womb with its promise of new life and the womb as tomb with its threat of death. With these inversions not only is 'the womb of the Great Mother . . . also a tomb' (Gilbert and Gubar 1989, p. 17), but the swamp is transformed from the womb of life into the jaws of death.

These inversions also involve seeing the swamp as a flow of menstrual blood, or perhaps more precisely albumen as Camille Paglia (1991, p. 92) argues that 'the primal swamp is choked with menstrual albumen, the lukewarm matrix of nature, teeming with algae and bacteria'. The swamp is a place of fecundity, which passively and supinely awaits what, in patriarchal and filiarchal terms, it lacks – the insemination of the phallic principle of reason, drainage, agriculture, canals, roads and the city to 'bring it to life'. In a discussion of Jean Renoir's film *Swamp Water* based on Bell's novel, Raymond Durgnat suggests that 'a swamp is a kind of flowing without issue, a too-swift growth decomposing while proliferating, a confusion of earth, water and vegetation, a seething with miasmas, suffocations and quiet murders' (Durgnat 1975, p. 231). Durgnat elaborates on what Gilbert Durand calls '*l'irrémédiable féminité de l'eau*' by demonstrating, as Durand puts it, that '*l'archetype de l'élément aquatique et néfaste est le sang menstruel*' (Durand 1969, p. 110).[12] This statement needs to be qualified by adding the rider that it is in patriarchal and filiarchal cultures that the archetypal aquatic and harmful element is menstrual blood. Menstrual blood is seen by patriarchy and filiarchy in negative terms as the flow which does not produce the issue of offspring. This flow is also seen as harmful which begs the question of harmful for whom? for what? Perhaps the flow is harmful for the patriarch because he does not bring it forth, because he does not cause it to flow unlike the flow of drains and canals.

In patriarchy and filiarchy fecund and flowing wetlands are the bad female genitals unlike the canalised and drained good female genitals (literally *la bonifica* in Italian, with a vulgar sexual reference but primarily 'land reclamation').[13] The menstrual flow defies the patriarch's, and the filiarch's, temporal control, his schedule and his calendar, as the swamp defies his spatial control, his maps and diagrams. In response to what he sees as an act of defiance to his law, the patriarch and the filiarch not only regards the menstrual flow with horror but denigrates it as producing something horrific, a monstrous thing, not living, not dead.

Without the insemination of the phallic, the menstrual flow is construed as, and in turn the swamp figured as, a grotesque parody of birth in which horrific and monstrous mutants are produced. Unlike zombies, these mutants decompose even as they proliferate. The menstrual flow for patriarchy and filiarchy is the ultimate swamp monster; the swamp, the quintessentially monstrous menstrual flow. The swamp may be, as Durgnat maintains, 'a state of Nature in terms of political philosophy', but it is also a state of monstrosity in terms of this culture (Durgnat 1975, p. 232).

In this tradition a journey into the swamp is inevitably couched in terms of penetration. In Bell's *Swamp Water* (Bell (1941) 1981) the narrator relates how 'they said Okefenokee held danger and unspeakable terrors; and yet for Ben it was a place of weird fascination. I'm a-going in her, he thought, "and it won't be long off"' (ibid. p. 6). When Ben goes into 'her' Okefenokee is figured as an obliging and sexually available woman: 'the swamp opened up and spread before them a wide, bonnet-covered, shallow-water prairie' (ibid. p. 92).[14] Playing on 'bonnet', the swamp is both woman and flower to be uncovered and deflowered when the swamp is inevitably 'penetrated' (ibid. p. 259).

The swamp is a secular hell, or underworld, through which the post-epic hero of the imperialist adventure story has to fight his way on his journey of sublimation and flight from the mother to manhood. According to Kant 'the loftiest flight that human genius made, in order to ascend to the sublime, took the form of adventures' (Kant 1960, pp. 114–15). Yet adventures were as Martin Green (1979, pp. xi and 3) puts it, 'the energizing myth of empire' and adventure stories 'the energizing myth of . . . imperialism'. The sublime is, in a word, imperialist. Yet in order to make the imperialist ascent into the heady heights of the sublime, the modern secular heavens, the hero of the imperialist adventure had to make the obligatory descent to the smelly depths of the slimy swamp as a post-epic, post-Christian underworld. Here the hero meets and defeats challenges, including swamps, and demonstrates his (advisedly) genius through his mastery of the tools, such as the compass and the gun, and techniques, such as writing a journal and drawing maps, of modernity (Green 1979, p. 23). For Kant, 'strivings and surmounted difficulties arouse admiration and belong to the sublime' (Kant 1960, p. 78), as they belong also to the adventure story and arouse admiration for its hero. In adventure stories, swamps are an obstacle to overcome, the depths to surmount, especially their smells (see Chapter 2, Figure 1).

The descent into the secular underworld of the swamp entails a return to the maternal (and matriarchal) womb of the wetland as home before returning to the paternal (patriarchal and filiarchal) home of the dry land, a simulation of the first home. In the words of Stratton-Porter's novel *Freckles*, the hero must make 'excursions into the interior', or perhaps more precisely incursions into the interior, journey through 'the dim recesses of the swamp', and finally descend 'into the depths of the swamp . . . through *what has been considered* impassable and *impenetrable* ways'

(Stratton-Porter (1904) 1990, pp. 32, 54 and 165, my emphases), through, in other words, the underworld. In patriarchal and filiarchal cultures the swamp is reduced to surface and the depths of the Body of the Mother denied through landscape painting and the map. The swamp is then penetrated by phallic heroes who invaginate it as the virgin (wet)land.

Only after penetrating the seemingly impenetrable can the hero then emerge triumphant with his manhood vindicated secure in the knowledge that he has sought out and subdued Mother Earth or Nature in her innermost recesses, that he has 'known' her in the sexual sense and so now her secrets hold no horrors for him because he has penetrated them. But by descending into the swamp as underworld, he runs a number of risks as Irigaray argues: 'the man who gets too close to the other risks merging with it. The man who stays too close to the other risks sinking into it. The man who penetrates the other risks foundering in it' (Irigaray 1991, p. 52). The man who gets too close to the swamp (m)other risks merging with it; the man who stays too long in the swamp (m)other risks sinking into it; the man who penetrates the swamp (m)other risks f(l)oundering in it. The swamp, as David Miller argues, is 'the underside of patriarchal culture, dominated by the body, materiality, corruption, infection, sexuality and irrationality – but also origin and creativity' (Miller 1989, p. 9). In patriarchal and filiarchal culture, the former are emphasised to the detriment, exclusion, and even repression of the latter; the body, human and environmental, have been associated with the former but not with the latter.

Notes

1. The parenthetical portmanteau 'hom(m)osexual' comes from Irigaray 1985, pp. 171, 172.
2. I am grateful to Ann-Marie Medcalf for pointing out the pertinence of Duras' work.
3. For national park forest as cathedral, see John Muir 1901, p. 80.
4. Davis, typewritten variant inserted at p. 13. The original reads: 'Dismal Swamp is alluring, capricious and contradictory. Like a beautiful woman she tantalizes with her oft changing moods. She encourages recklessness, but metes out swift and cruel punishment to those whom she lures into her web of dangers.'

5. According to the lurid biographer of the young, slighted Robert Frost, 'it may have crossed his mind that he could secretly arrange death by plunging into the very heart of some such Slough of Despond as the Dismal Swamp in Virginia – a treacherous, dangerous, frightful, and snake-infested wilderness' (Thompson 1966, pp. 176, 177).

6. I am grateful to Jim Warren for pointing out the pertinence of this book.

7. 'Kidneys', *Shorter Oxford English Dictionary*, p. 1153.

8. I am grateful to Wendy Parkins for drawing my attention to this book.

9. For Lyotard (1993, p. 250), 'the theoretical is a major procedure of invagination'. The theoretical sublime invaginates the surface of the earth and interiorises its depths in an inversion of its spatial arrangement. Perhaps Figure 1 in Chapter 2 should be turned upside down!

10. See also Showalter 1991, p. 86. I am grateful to Wendy Parkins and Jon Stratton for drawing my attention to Showalter, who quotes Gilbert and Gubar.

11. For the tomb as womb, see Gimbutas 1989, pp. 151–9. See also Sjöö and Mor 1991, pp. 8, 46, 51, 52, 59, 76, 103, 113, 180 and 184.

12. I am grateful to Peter Gilet for pointing out the pertinence of this book.

13. For a descriptive, uncritical discussion of *lu bonifica* as Italian wetland reclamation, see Harris 1957, pp. 308–15.

14. Sesonske (1982, p. 53) maintains that in Renoir's film Ben going into Okefenokee is filmed as a 'penetration into the open but entangled vastness of the swamp'.

IV

Minds and Marshes

7

The Melancholic Marshes and the Slough of Despond: The Psycho(eco)logy of Swamps

The wetland has not only been seen as bad for the body in the patriarchal tradition, but also as bad for the mind. Indeed, it can plunge the mind into melancholia, and even into madness. In a typical assessment made one hundred years ago, the later-to-be President of the United States Theodore Roosevelt (1893, p. xvi) referred to 'the melancholy marshes' in the 'lonely lands' of 'the wilderness'. More specifically, the corpse beneath the swamp is not only a reminder of death, but also a vector of melancholy. Such sentiments were voiced when Hubert Davis referred to 'the stench of a rotting carcass of some depraved hunter lost in the depths of its [the Great Dismal Swamp's] melancholy jungles' (Davis 1962, p. 21).

This assessment of the marsh as melancholic can also be found in the work of those masters of dejection, the Romantic poets. Percy Bysshe Shelley, for example, referred to 'a wide and melancholy waste/Of putrid marshes' when evoking 'the Spirit of Solitude' (Shelley 1975, pp. 51–2). Previously Samuel Pepys in the late seventeenth century travelled 'over most sad Fenns (all the way observing the sad life that the people of that place . . . do live)' (Pepys 1972, p. 311). More recently, the map of 'the world' of those 1920s classics of Anglophone children's literature, *Winnie-the-Pooh* and *The House at Pooh Corner*, with its Christopher Robin of the clean and proper body, has that arch manic-depressive donkey Eeyore living in 'Eeyore's Gloomy Place' which is 'rather boggy and sad' (Milne 1958, frontispiece map).[1]

Even more recently, the three central characters in Marguerite Duras' 1950 novel *The Sea Wall* (*Un barrage contre le Pacifique*),

155

set in the Mekong Delta in which they have 'their isolated piece of the salt-soaked plain', are, in turn, 'soaked in boredom and bitterness'. This wetland was 'a wasteland where nothing would grow' in which there lived 'wasteland [rather than wetland] people' inflicted with 'the same old deadly solitude' (Duras 1986, p. 9). In Duras' rewriting of *The Sea Wall* over thirty years later as *The Lover* (*L'Amant*), set, like its predecessor, in 'the biggest deltas in the world . . . made of black slime,' the wasteland is also a wilderness, even 'the salt land's a wilderness too' (Duras 1985, pp. 28, 92).

The association between wetlands and melancholy (and boredom and bitterness, sadness and solitude) has its roots in the theory of the elements and the humours. Patriarchal European culture was founded and still functions on the philosophy of the four elements: earth, air, fire and water. Wetlands have always been problematic, indeed aberrant, from this point of view because they mix the elements of earth and water (and even air and 'fire' or heat in the tropics). They also cross the boundaries between land and water, and can even be in transition spatially and temporally between open water to dry land. They are a troubling and unsettling category mediating between land and water.

As wetlands mix the elements, they produce an aberrant 'humour', or psychosomatic state, strictly a kind of phlegmatic melancholy. In the patriarchal western tradition wetlands have been seen as a wilderness to be tamed, the sites of mixed elements and aberrant humours giving rise to melancholy and madness. In this chapter I trace, deconstruct and decolonise the association between melancholy (and depression, despair, despondency, dread and the dismal) and wetlands via an initial consideration of the theory of the elements and the humours.

THE ELEMENTS AND THE HUMOURS

The theory of the elements was developed by the pre-Socratics and later enunciated by Aristotle in the treatise which came to be called 'De Generatione et Corruptione' ('On Coming-to-be and Passing-Away'). For Aristotle what he called the four 'so called elements' (Ross 1995, p. 104)[2] of earth, air, fire and water were produced by the selective 'coupling' of the four 'elementary qualities' and two 'contrarities' of hot, cold, dry and wet (or 'moist', or 'fluid' as one commentator prefers) (Ross 1995, p. 105). As a result earth is cold and dry, air hot and moist, fire hot and dry, and water cold and moist. Theoretically it was possible for the four 'elementary qualities' and two 'contrarities' of hot, cold, dry and wet to be combined into six couples, but as it is impossible for something to be both hot and cold, wet and dry (though not in practice for a solid to be damp or sodden as in a slimy wetland, nor for a solid to be sublim(at)ed into a gas by heat), there were in effect for Aristotle only four proper couples. It was thus acceptable for the elementary qualities to mix, but only in four prescribed or proper ways to produce the four elements. The mixing of the 'contrarieties' of hot and cold, wet and dry was 'impossible', or in David Ross' terms they refuse to be coupled though Aristotle later concedes that 'all the 'elements' will be able to be transformed out of one another' (Aristotle, GC329a–332b) which would involve coupling of the contrarities.

Wetlands were thus effectively ruled out of court in this cosmogony on two counts depending on whether they were considered at the level of elements or qualities. They were ruled out at the level of elements because they combine earth and water, or because they mediate between earth and water spatially or temporally. They were also ruled out at the level of elementary qualities for as the earth is cold and dry, and water cold and wet, a mixture of or a mediating category between the two sets of contrarieties would produce the quality of cold, dry and wet, an evident impossibility in this philosophical schema. The sublime is even more out of court as it involves a transition through the elements from 'earth' (or solid) to 'air' (or gas) by or via 'fire' (or heat). It would also entail the absurd combination of the qualities of hot, cold, dry and wet.

Aristotle distinguishes the elementary qualities of the hot and cold from the wet and dry in terms of the active/passive distinction as 'the first pair implies *power to act* and the second pair *susceptibility*' (GC329b). Or as David Ross put it, 'hot-cold play in general the part of agent and dry-fluid that of patient'. The distinction between agent and patient is gendered in patriarchy and filiarchy. The former is associated with virile masculine activity and the latter with supine feminine passivity. Aristotelian philosophy is no exception as it, according to Carolyn Merchant, 'associated activity with maleness and passivity with femaleness' (Merchant 1980, p. 13). On the Aristotelian view of (re)production the masculine hot-cold inseminates the feminine wet-dry and brings forth the four elements. A gender distinction also applies to the elements themselves. Of the four elements for Aristotle 'Fire and Earth . . . are extremes and purest' and 'Water and Air . . . are intermediaries and more like blends' (GC330b–331a). In other words, fire and earth are definite and discrete, and thus 'masculine'; water and air are mediatory and mixed, and therefore 'feminine'. Fire and earth are privileged over air and water, hot and cold over wet and dry so the figure below is hierarchical.

Yet Aristotle accepted that 'Earth has no power of cohesion without the moist' (presumably just as long as the moist stayed between the bits of earth and did not mix with them thus dampening them) (GC335a). The 'feminine' moist (the wetlands), holds the 'masculine' earth (the dry lands) together; without the former the latter would fall apart. Wetlands are, as a recent article argues, 'the glue that holds the land together', or more precisely, the soft glue that holds, in the words of Davison's *Soundings*, 'the hard bones of the land' together (Mitchell 1992, p. 15; Davison 1993, p. 85). The moist, moreover for Aristotle, is 'that which being readily adaptable in shape is not determined by any limit of its own' (GC329b). This 'feminine' adaptability and limitlessness of water is the condition of possibility for modern 'masculine' waterworks.

Yet whilst Aristotle was appreciative of the cohesive and other qualities of the moist, he certainly did not like the inundatory cr penetrative qualities of the damp and the sodden which result from the solid and the liquid, earth and water, the dry and the wet, the masculine and the feminine 'impossibly' combining in aberrant ways: '"damp" is that which has foreign moisture on its surface ("sodden" being that which is penetrated to its core)' (GC330a). A wetland in these terms is earth with 'foreign' moisture on it, as in a lagoon, or earth 'penetrated' with moisture to its core, as in a marsh. Either way the implication seems to be that the 'masculine' earth should not have 'feminine' water lying on its surface or penetrating its depths. The only proper function for 'feminine' water is to make the 'masculine' earth stick together. On this view and by implication, earth should not have the 'foreign' matter of water on it, nor should it be 'penetrated' by the invasive power of water. Earth should be dry.

Milton followed Aristotle's account of the elements by only admitting the possibility of the mixture of four elementary qualities into what he called 'the cumbrous elements, earth, flood, air, fire' (*Paradise Lost*, III, 715):

> ye elements, the eldest birth
> Of Nature's womb, that in quaternion run
> Perpetual circle, multiform, and mix
> And nourish all things, let your ceaseless change
> Vary to our great Maker still new praise.
> > (*Paradise Lost*, V, 180–4)

The mixing of the elementary qualities in Nature's womb does nourish all things, but not only the four mixtures into the elements that Aristotle and Milton allowed. Indeed, the two aberrant mixtures of wet and dry, hot and cold produce the temperate and tropical wetland which are far more productive than both dry land and cold water. It is the wet land and warm water which give birth to life, which are the womb of the world.

For Aristotle the arrangement of the elementary qualities and their proper couples were associated with physical conditions and with mental humours or temperament. In his treatise 'De Problemata' in discussing 'problems connected with the effect of locality on temperament,' Aristotle goes on to answer with another question the question he poses:

> why is it that those who live in airy regions grow old slowly, but those who inhabit hollow and marshy districts age quickly? Is it because old age is a process of putrefaction, and that which is at rest putrefies, but that which is in motion is either quite free from, or at any rate less liable to, putrefaction, as we see in water? In lofty regions, therefore, owing to the free access of the breezes, the air is in motion, but in hollow districts it stagnates. Furthermore, in the former, owing to its movement, the air is always pure and constantly renewed, but in marshy districts it is stagnant. (Aristotle, Probl. 909b)

Not only is the water stagnant in marshy districts, but the air as well. No wonder such districts are bad for you both physically and mentally. Despite the differences between his work and that of his teacher, Aristotle reproduces, like a good pupil, Plato's Socrates' denigration of 'hollow places' which we saw evidenced in the *Phaedo* in Chapter 2.

In the western theory of the elements, the four elements combine to form or are associated with psychosomatic states or 'humours' as developed by the Elizabethans and as formulated by E. M. W. Tillyard (1963, p. 76). These states can be tabulated in terms of the mixing of the four elementary qualities into the four legitimate elements, in turn into four more modern qualitative elements and finally into four proper substances and humours:

	HOT	COLD
DRY	FIRE HEAT CHOLER/IC	EARTH SOLID MELANCHOLY/IC
WET	AIR GASEOUS BLOOD/SANGUINOUS	WATER LIQUID PHLEGM/ATIC

This schema becomes problematic when considering the association between melancholia and wetlands for in the terms of this schema they should not strictly be associated. Or more precisely, wetlands as between or combining earth and water should be seen as equally phlegmatic as they should be seen as melancholic. As a result this combination of elements would produce a phlegmatic melancholy, an aberrant humour.

Traditionally, however, indigenous wetlanders have been seen as either melancholic or phlegmatic, or tending from the latter to the former. In Graham Swift's *Waterland* fenlanders are phlegmatic people with a phlegmatic humour and a phlegmatic sense of humour, 'a muddy, silty humour' (Swift 1984, pp. 13, 14). Phlegm, 'or slime,' for Crick the narrator, like Sartre the philosopher, is 'an ambiguous substance. Neither liquid nor solid: a viscous semifluid . . . It deters the sanguine and the choleric and inclines towards melancholy' (ibid. p. 298). The phlegmatic is not stable in the table of elements and humours. So much does the phlegmatic incline towards the melancholic in the Fens that it can produce 'poor Fenland madmen and melancholics' (ibid. p. 82). The Fens draw out an 'inward melancholy' and can produce 'muddy madness' (ibid. pp. 186 and 190).

Even though the melancholic has traditionally been associated with the cold and dry, a visit to the cold and wet/dry marshes by the drylander can induce melancholy, whereas the indigenous wetlander by temperament is usually phlegmatic, but tends towards the melancholic. By temperament, the wetlander at home in the wetland is phlegmatic (associated with the watery), whereas the drylander visiting the wetland tends to be infected, or more precisely affected, by the melancholic (the earthy, though more precisely the earthy in combination with the watery).

One of the 'causes of melancholy' for Robert Burton writing in the early seventeenth century in his *The Anatomy of Melancholy* was 'standing waters'. He went on to elaborate in Hippocratic terms that:

standing waters, thick and ill-coloured, such as come forth of pools and moat where hemp is steeped or slimy fishes live, are most unwholesome, putrefied, and full of mites, creepers, slimy, muddy, unclean, corrupt, impure, by reason of the sun's heat and still standing; they cause foul distemperatures in the body and mind of man, are unfit to make drink of, to dress meat with, or to be used inwardly or outwardly ... So that they that use filthy, standing, ill-coloured, thick, muddy water, must needs have muddy, ill-coloured, impure and infirm bodies. And because the body works upon the mind, they shall have grosser understandings, dull, foggy, melancholy spirits, and be readily subject to all manner of infirmities. (Burton 1932, pp. 224–5)

Although Burton enunciates a psychosomatic theory of the body and mind, he subscribes to a simplistic analogy, or even corre-spondence theory, based primarily on colour, between the natural environment and the human body. And although he does not subscribe to the miasmatic theory of physical illness, it is the standing waters themselves, rather than the vapours which rise from them, which are to blame for illness, both mental and physical.

Not only can standing waters cause melancholy, so also can the cold and dry air which rises from them. When Burton discusses 'bad air', the worst air of all out of hot and cold air, he describes it as 'a thick, cloudy, misty, foggy air, or such as come from fens, moorish grounds, lakes, muckhills, draughts, sinks, where any carcasses or carrion lies, or from whence any stinking fulsome smell comes: . . . such air is unwholesome, and engenders melancholy, plagues and what not' (ibid. p. 239). It is on such an 'anatomy' or miasmatic theory of melancholy and malaria that the Sanitary Movement of two centuries later was founded.

Yet such air is surely not dry but wet. Under the traditional western theory of the elements and humours as formulated systematically by the Elizabethans, the wetland did not fit easily into the schema as we have already seen. In fact, melancholy was strictly associated with the dry upland of the agricultural country and the modern city, even with capitalism. Eagleton argues that 'melancholia is an appropriate neurosis for a profit-based society' (Eagleton 1986, p. 41), for modern sublimated culture. Further, Kant postulated that 'the feeling of the sublime is 'sometimes accompanied with a certain dread or melancholy' (Kant 1960, p. 47).

In response to the modern melancholic city sublime, there was a massive attempt to displace blame onto the wetland. The Elizabethan 'world picture' as enunciated by E.M.W. Tillyard was not flattering to the modern melancholic and sublimated city-dwelling drylanders; it was more flattering to the pre-modern phlegmatic and 'slimated' fen-dwelling wetlanders. Wetlands did not 'fit the picture', but rather than changing the picture, it was the wetland that had to be changed, or more precisely in Burton's case the dwellers in wetlands, including in cities built on or near wetlands (invariably 'foreign'), denigrated.

After enumerating a list of places where the air is bad, Burton concludes by asking 'how can they be excused that have a delicious seat, a pleasant air, and all that nature can afford, and yet through their own nastiness and sluttishness, immund and sordid manner of life, suffer their air to putrefy, and themselves to be choked up?' (Burton 1932, p. 240) The wetland, its standing waters, its bad (but wet) airs, and its filthy, sexually loose and morally reprehensible dwellers, either indigenous wetlanders or city-in-wetland dwellers, is the cause of melancholy displaced from the cold and dry uplands.

THE GREAT DISMAL SWAMP

Try as they might, travellers into the Great Dismal Swamp had enormous difficulty in preventing what they saw as the steaming hot and dripping wet horror of the place from 'oppressing their spirits' or depressing them. By its very name the Great Dismal Swamp seems to conform to the European idea of the wetland associated with the dreary and the dreadful.

When David Hunter Strother, better known under his pen-name of 'Porte Crayon', travelled into the Great Dismal Swamp in 1856 he made a journey not only into a region which had haunted his imagination since his 'earliest recollection', and even preceded it so associated with birth and the womb, but also into hell (so associated with death, the tomb and the end of life, rather than into death, heaven and the beginning of new life) (Strother 1959, p. 133). Strother's journey is consequently more into the womb as tomb rather than into the tomb as womb:

lofty trees threw their arching limbs over the canal, clothed to their tops with a gauze-like drapery of tangled vines; walls of matted reeds closed up the view on either side, while thickets of myrtle, green briar, bay, and juniper, hung over the black, narrow canal, until the boat could scarcely find a passage between. The sky was obscured with leaden colored clouds, and all nature was silent, monotonous, deathlike. The surface of the canal was glassy smooth, and reflected the towering trees, the festooned vines, and pendant mass, with the clearness of a mirror. (ibid. pp. 133–4)

Strother's journey into the swamp, like that depicted in Conrad's 'Lagoon' and Haggard's *She*, is initially through a narrow birth canal. Strother's journey is also a journey into a place of inverted mirror images where the reflections in the water reproduce the scene above the water and make it doubly oppressive and depressive. Strother's journey also takes place beneath the black sun of melancholia. The Great Dismal Swamp is doubly melancholic because Strother finds that the 'leaden coloured clouds' and its claustrophobic and suffocating vegetation are reflected in the stagnant waters producing an inverted image of itself.

This is a place of dreary stillness where Strother thinks that 'I was alone, utterly alone', though noting in the very next sentence the presence of 'my men' (ibid. p. 134). But they were not really men as they 'might have been ghouls, or cunningly-devised machines, set in motion by some malignant sorcerer, to bear me away into a region of stagnation and death' with its 'pools of black, slimy water' (ibid. pp. 134–5). These ghoul-machines, these swamp monsters, like dolls and automata for Freud, produce a sense of the uncanny, a return to the repressed of death amongst other things (see Chapter 2, Figure 1). Not only is the swamp described as the region of death, but also of 'monotony [which] is wearisome, dreary, solemn, terrible' (ibid. p. 135).

The swamp with its (birth) canal gives onto the womb of Lake Drummond, 'a broad sheet of water' (unlike the narrow birth canal) at the centre of the Great Dismal Swamp (ibid. p. 136). The lake fulfils all his childhood dreams and so 'it was neither new nor strange' (ibid. p. 137). Strother has dreamt his own birth and here he re-enacts it in reverse like Kerans in Ballard's *The Drowned World*. Rather than the uncanny and unhomely experience of the swampy birth canal, the Lake is canny and homely, though its waters are 'dark' and 'leaden' like the black sun of melancholy over the canal (Strother 1959, p. 137). The black waters, indeed, reflect the black sky. It is even a region of monstrosity and death with its 'gigantic skeletons of cypress' festooned with 'weepers of funereal moss' serving to produce 'a picture of desolation – Desolation' (ibid. p. 137). The Lake is an ambiguous mixture of life and death, of the new and strange, and the old and familiar, of the uncanny and canny.

Like the Mekong Delta in Duras' *The Lover*, the Lake was for Strother a 'vast wilderness' and 'a gloomy and inaccessible retreat' (Strother 1959, p. 141). At night it is even worse for the Lake becomes 'a wild, weird scene, suggestive to the imagination of more than language can express' (ibid. p. 141). By day the Lake is canny, but at night it is uncanny. Like the sublime which it doubles, or of which it is the obverse, the slimy is on the margins of language, like the wetland is on the edge of consciousness and 'civilisation'. Strother, like Sartre, struggles to find the words to put the slimy into language. The slimy defies language and his attempts to encapsulate it, and so all he can do is to revert to the almost cliched, certainly stereotypical, images of the swamp as imaginary hell.

Strother subscribes to a miasmatic theory of mental illness, or at least of loneliness and monotony, though he does not subscribe to a miasmatic theory of physical illness. In fact, there is a strong contrast in his account of the Great Dismal Swamp and Lake Drummond between his description of the mental horrors produced by the place and his reporting that 'the interior of the Swamp is perfectly healthy' as its waters are 'healthful' and as the place itself is 'free from miasmatic diseases' (ibid. pp. 143, 144). There is also a strong contrast between his horror of the constricting and constraining birth canal through the swamp and his delight in the open womb and broad waters of daytime Lake Drummond, and so a strong contrast between the Lake/womb by day as place of light (and even of life) and by night as place of darkness (and even death).

Darkness and the sun, which is the quintessential fire in the European cosmogony, play a crucial part in the aetiology of the dismal and melancholy, especially in conjunction with, or more precisely produced by clouds rising from, the wetland. In Rider Haggard's *She* 'the sun ... drew thin sickly looking clouds of poisonous vapour from the surface of the marsh and from the scummy pools of stagnant water' (Haggard (1887) 1991, p. 63). These clouds do not so much produce physical illness as mental oppression by blotting out the sun. Wetlands 'cause' or produce phlegmatic melancholy not only because they mix earth and water, but also because their vapours can blot out the sun (or the element of fire), though black clouds in general (and not specifically associated with wetlands) have been seen as productive of melancholy, as, for example, in Hoffmann's 'Sandman'.

For Freud 'a phantasy of returning into his mother's womb', is 'the substitute for copulation' for 'a man who is impotent (that is, who is inhibited by the threat of castration)' (Freud 1979, p. 296). So all those heroic men who descend into the underworld of the swamp as a phantasy of returning to their mother's woman are impotent, or inhibited by threatened castration. This phantasy can be given rise to by anxiety about castration, and not only one's own. For Freud, 'the first experience of anxiety which an individual goes through is birth, and, objectively speaking, birth is a separation from the mother. It could be compared to a castration of the mother (by equating the child with a penis)' (ibid. p. 286; see also pp. 233 and 245). The phantasy of a return to the mother's womb would then be the substitute for copulation for the man who is inhibited by the 'castration,' or sense of loss, of the mother.

This sense of loss is not only experienced by the anxious man in relation to his mother, but also in relation to his own turds. Freud argues that 'castration can be pictured on the basis of the daily experience of the faeces being separated from the body or on the basis of losing the mother's breast at weaning' (ibid. p. 285). Patriarchal and filiarchal men give 'birth' to faeces, their hard objects which they fear losing as they have lost the soft object of the breast, and they become anxious when their turds become soft and slimy, when they cease to be proper objects to be mastered and become improper abjects. Indeed, for Freud 'the loss of object' is 'a determinant of anxiety' (ibid. p. 295; see also p. 325). In other words, loss of object denotes in his vocabulary 'castrated,' or in Kristeva's vocabulary abject. The phantasy of a marsh monster, as we will see in the following chapter, is the substitute for copulation for the oral-sadistic man who fears castration, or more precisely fears that the oral sadism which he directs against the mother and mother earth will be exercised back upon him, or upon his 'precious portion' as Freud calls it, in an act of symmetrical reciprocity (Giblett 1993, pp. 541–59).

THE SLOUGH OF DESPOND AND THE GRIM DOMAINS OF GIANT DESPAIR

The aberrant mixture of the elements of earth and water in the temperate wetland – as well as heat in the tropical – produces the humour of the phlegmatic, tending to the melancholic, if not to downright despondency and despair. Rider Haggard's *She* set in 'deepest, darkest Africa' makes repeated reference to 'these dreadful marshes,' 'that dreary marsh', 'that dreadful wilderness of swamp' and to 'the dismal swamp' (pp. 61, 72, 73 and 113). Similarly Charles Kingsley describes the Fens with its 'dreary mud-creek . . . and dreary flats' as 'a lonely land' (Kingsley 1908, pp. 229, 230).

Yet the most famous depiction in English culture of this association is, of course, John Bunyan's 'Slough of Despond' in *Pilgrim's Progress*. Louis Marin has described *Pilgrim's Progress* as 'the most influential religious book ever composed in the English language' (Fritzell 1978, p. 528). It is also the second most published book in the English language after the Bible. No doubt part of its pre-eminent influence has been to educate generations of readers that an ecologically functioning wetland is an ellegor-ical emblem for a sump of iniquity that would drag the unsuspect-ing and unwary Christian down and entrap him/her for all eternity unless they are 'saved': 'this miry slough is such a place as cannot be mended; it is the descent whither the scum and filth that attends conviction for sin doth continually run, and therefore it is called the Slough of Despond'.

The same figure is used with the same pejorative overtones (though not of course as an ellegorical emblem for sin) in Donna Haraway's recent proclamation that 'like Christian in *Pilgrim's Progress* . . . I am committed to skirting the slough of despond and the parasite-infested swamps of nowhere to reach more salubrious environs' (Haraway 1992, p. 295 and p. 329, n. 1). One wonders salubrious for whom? In whose terms? And if the slough of despond is not an allegorical emblem for sin, what is 'sin' in Haraway's theologised, or at least moralised, wetlandscape? What is 'nowhere'? Is it 'the amniotic effluvia of terminal industrialism' that Haraway refers to later (but without making any explicit connection) which I have suggested earlier is parasitic on pla-cental wetlands past and present? If so, then the figure would have a critical edge and contemporary pertinence; if not, then Har-away's use of the figure would seem to be gratuitously uncritical about the use of tropes and blind to the politics of place and to the placental functions of wetland (place-blind and even misaqua-terrist).

Yet Haraway argues in these very same two pages that 'nature is . . . a *topos*, a place,' that 'nature is also *trópos*, a trope' and later for a 'politics of articulation' which speaks with an intersubject rather than for 'a politics of representation' which speaks for and on behalf of an object (ibid. pp. 296 and 311–13). But as nature is also places, *topoi*, as different *topoi* are troped in diffrent ways with some being valorised at the expense or to the detriment of others, so there is a politics of the articulation of tropes of topes, a politics of *trópoi* of *topoi*, not least of wetlands. Haraway cares about 'the survival of jaguars and the chimpanzee, and the Hawaiian land snails, and the spotted owl, and a lot of other earth*lings*, (ibid. p. 311) but does she care about the survival of sloughs (of despond), (parasite-infested) swamps, uncharismatic microfauna (like parasites) and placental wetlands?

No such qualms of conscience or ethical dilemmas were to trouble English writers after Bunyan despite the increasing secularisation of English culture. The Slough of Despond was divested of its religious overtones, but not of its pejorative, misaquaterrist associations. Instead of being a place of evil construed in religious terms, it became a place of melancholia, a kind of secular despondency or despair. Dickens, for example, has 'the Slough of Despond' in *Hard Times*, as well as 'the slough of inanity'. If a working-class room could be troped as an African Swamp by Southwood Smith, 'the social swamp' as Thomas Huxley called '*la misère*' could be 'a Slough of Despond' (Huxley 1989, p. 335).

Dickens, however, saves his most scathing attack on the swamp as a Slough of Despond for Chapter 23 of *Martin Chuzzlewit*. In an earlier chapter we saw how Dickens subscribed in this novel to the miasmatic theory of malaria; he also subscribed to a miasmatic theory of melancholia which reproduces the Aristotelian and Elizabethan theory of the elements and the humours. The final leg of the journey to Eden at the confluence of the Mississippi and Ohio Rivers is through a scene made up entirely of

sky, wood and water, all the livelong day; and heat that blistered everything it touched. On they toiled through the great solitudes, where the trees upon the banks grew thick and close; and floated in the stream; and held up shrivelled arms from out the river's depths; and slide down the margin of the land, half growing, half decaying in the miry water. On through the weary day and melancholy night: beneath the burning sun, and in the mist and vapour of the evening: on, until return appeared impossible, and restoration to their home a miserable dream. (Dickens (1844) 1953, p. 367)

This is a journey into a place where the elements are mixed, a place which, as a result, produces 'the dull depression of the scene'. Like Strother's journey into the Great Dismal Swamp, the journey is a secularised epic descent into the underworld on 'old Charon's boat, conveying melancholy shades to judgment' (ibid.). Judgement of whom and for what is not specified, but the post-epic journey through the swamp as modern secular hell is part of the test which Martin must complete to achieve manhood on his journey to return home.

From this point the situation can only deteriorate as these are only the approaches to the underworld; this is not Hell itself but only the journey across the River Styx:

as they proceeded further on their track, and came more and more towards their journey's end, the monotonous desolation of the scene increased to that degree, that for any redeeming feature it presented to their eyes, they might have entered, in the body, on the grim domains of Giant Despair. A flat morass, bestrewn with fallen timber; a marsh on which the good growth of the earth seemed to have been wrecked and cast away, that from its decomposing ashes vile and ugly things might rise; where the very trees took the aspect of huge weeds, begotten of the slime from which they sprung, seeking whom they might infect, came forth at night, in misty shapes, and creeping out upon the water, hunted like spectres until day; where even the blessed sun, shining down on festering elements of corruption and disease, became a horror; this was the realm of Hope through which they moved. (ibid. p. 369)

For Dickens Eden is ironically not a paradise but a festering sink in which the elements are improperly mixed and from this improper mixing spring hideous and monstrous deformities and mutants. Yet by a further countervailing irony the wetland chaos, as I have suggested in the previous chapter, could be seen to be God's first work, not so much Eden, the pastoral paradise, but the pristine pre-Edenic, pre-pastoral and pre-paradisaical wetland. Dickens' Eden, however, is far from that watery chaos. It is both a 'hideous swamp' 'choked with slime' and also a 'dreary situation' which produces 'dread' and 'the apprehension of death'. In the approaches to the swamptown the elements of heat, water, air and earth are mixed; in Eden itself they are also mixed where 'a fetid vapour, hot and sickening as the breath of an oven, rose up from the earth, and hung on everything around'.

Like the slime for Sartre and the swamp for Byrd on which they could not leave a mark and which they found horrific because it threatened to leave a trace on them, so too at Eden Martin could not leave his footprints on 'the marshy ground' for as soon as his foot sank into it, 'a black ooze started forth to blot them out'. He is also unable to leave much of a mark on the land, or more precisely wetland, as it was 'mere forest' whose trees had grown so thick and close that they shouldered one another out of their places, and the weakest, forced into shapes of strange distortion, languished like cripples. The best were stunted, from the pressure and want of room; and high about the stems of all, grew long rank grass, dank weeds, and frowsy underwood: not devisable into their separate kinds, but tangled all together in a heap; a jungle deep and dark, with neither earth nor water at its roots, but putrid matter, formed of the pulpy offal of the two, and of their own corruption. (ibid. p. 372)

This wetland is a nightmare for the taxonomising botanist who would seek to separate out its specimens of flower, fruit and leaf into genus and species. Here is not just the excrement of earth and water, but the corrupt, decomposing excrement of earth and water.

The Slough of Despond, rather than being rehabilitated or its pejorative connotations reversed in the process of increasing secularisation, came to stand for the dark side of the Enlightenment and Romanticism (and even Victorianism both British and American as Dickens' and Strother's work attests), despite their differences. It is, as Peter Fritzell has argued:

the well-remembered antithesis to enlightenment and romantic thought . . . With the declaration of man's [sic] perfectibility and the affirmation of sublime and picturesque vistas, with the denial of man's imperfectibility and the negation of landscapes which do not fit the conventions of the sublime or the picturesque people of the nineteenth century can proceed to their new-found manifest destiny. They can proceed with the story of exploitation, the story of draining, ditching, clearing and filling, which will improve, civilise, humanise, and finally redeem the nonhuman environment. (Fritzell 1978, p. 528)

Such a place as the Slough of Despond as the antithesis of enlightenment would seem to make the ideal setting or backdrop for the private detective story since it is a place of almost impenetrable evil and darkness which can only be pierced by the penetrating light of reason brought to bear upon it by the superior intellect and insight of the great man himself (for it is invariably a man). It is hardly surprising then that the most famous fictive private detective of them all in probably his most famous story should have to deal with a Slough of Despond. This is, of course, Sherlock Holmes in *The Hound of the Baskervilles* with its Great Grimpen Mire.[5]

Dr Watson, taken on a tour of the moor by Stapleton, is asked if he notices 'those bright green spots scattered thickly over it'. Watson replies, 'yes, they seem more fertile than the rest'. His guide laughs and goes on to say 'that is the Great Grimpen Mire' which he then describes as 'the bog-hole' and as 'an awful place'. By recognising the fertility of the mire, Watson was a wetlands ecologist before his time. Whilst the Mire is being discussed 'a long, low moan, indescribably sad swept over the moor. It filled the whole air, and yet it was impossible to say whence it came. From a dull murmur it swelled into a deep roar, and then sank back into a melancholy, throbbing murmur once again'. Watson asks what the sound is to be told that 'the peasants say it is the Hound of the Baskervilles calling for its prey'. Not satisfied with this answer, dismissing it as 'nonsense' he presses Stapleton for a more rational explanation which he feebly offers:

'bogs make queer noises sometimes. It's the mud settling, or the water rising, or something'. 'No, no that was a living voice'. 'Well, perhaps it was. Did you ever hear a bittern booming?' 'No, I never did'. 'It's a very rare bird – practically extinct – in England now, but all things are possible upon the moor . . .' 'It's the weirdest, strangest thing that ever I heard in my life'. 'Yes, it's rather an uncanny place altogether'.

Indeed, as I have suggested earlier, the wetland is the uncanny place par excellence for patriarchal western culture.

Watson is greatly affected by 'the melancholy of the moor' and reports to Holmes from what he calls 'this most God-forsaken corner of the world. The longer one stays here the more does the spirit of the moor sink into one's soul, its vastness, and also its grim charm. When once you are out upon its bosom you have left all traces of modern England behind you'. This conclusion is hardly surprising because once you have entered the mire mother you have entered a pre-modern, even pre-Christian, landscape, or more precisely wetlandscape, of mother earth the precursor of the modern landscape and the 'foundation' for it. Even though one may avoid sinking into it physically, it sinks into one metaphysically and psychologically. It is not a settled and stable object, but its own free agent, a dynamic process which cannot be kept within strict bounds, whose primary horror is that it breaks boundaries.

In the conclusion to the story Holmes and Watson follow Stapleton into 'the widespread bog' with its 'green-scummed pits and foul quagmires' where

rank weeds and lush, slimy water-plants sent an odour of decay and a heavy miasmatic odour onto our faces, while a false step plunged us more than once thigh-deep into the dark, quivering mire . . . Its tenacious grip plucked at our heels as we walked, and when we sank into it it was if some malignant hand was tugging us down into those obscene depths, so grim and purposeful was the clutch in which it held us.

Although Watson as a modern medical man does not seem to subscribe to a miasmatic theory of malaria, he supports a miasmatic theory of melancholia. He also anthropomorphises the mire by giving it hands to pull both him and Holmes down into its clutching, (s)mothering embrace as it had pulled Stapleton 'down in the foul slime of the huge morass' though, of course, Holmes and Watson as enlightenment heroes escape its clutches and emerge triumphant the crimes, and riddle, of the mire solved and their rationalist manhood vindicated.

THE MELANCHOLIC TARN

Probably the writer to have most thoroughly explored and exploited the conjunction between the aquaterrestrial and the melancholic is Edgar Allan Poe, especially in *The Fall of the House of Usher* with its 'black and lurid tarn' which Camille Paglia describes somewhat luridly as 'the primeval swamp of generation . . . Its stagnant morass is mother nature's mouth in fetid repose', or perhaps more precisely mother nature's *vagina dentata* in patriarchal projection (Paglia 1991, p. 576). The first-person narrator/character of this Gothic tale rides to 'the melancholy House of Usher' through 'a singularly dreary tract of country' which has also been described by one critic as 'a gripping Gothic ruin of a landscape' (Pollin 1983, p. 252). Not only Gothic in its physical characteristics, the landscape is also Gothic in terms of its mental and physical affects, particularly its play on fears of engulfment and entrapment. It seems to have no quality to redeem it or to lift 'an utter depression of soul' which befalls the narrator.

In fact, this tract of country produces 'an iciness, a sinking, a sickening of the heart – an unredeemed dreariness of thought which no goading of the imagination could torture into aught of the sublime . . . that half-pleasurable, because poetic, sentiment with which the mind usually receives even the sternest natural images of the desolate or terrible'. The only sentiment with which the mind could receive these images was the obverse of the sublime, what Julia Kristeva has called 'the slimy poison of depression' (Kristeva 1991, p. 10). This poison 'reeked up from the decayed trees' and from 'the sullen waters of the tarn' producing 'a pestilent and mystic vapour, dull, sluggish, faintly discernible, and leaden-hued' like the clouds of the Great Dismal Swamp for Strother. Again, as in his journey the 'decayed trees' are reduplicated by being reflected in 'the still waters of the tarn' which only serves to heighten the sense of decay by doubling it and inverting the images of 'the ghastly tree-stems' in an upside-down world, the world of the House of Usher.

The pestilent and mystic vapour seems to invade the House as well. Outside there was 'an atmosphere which had no affinity with the air of heaven;' in other words, it only had affinity with the air of hell, the swamp as hell. Similarly inside the house there was 'an atmosphere of sorrow. An air of stern, deep and irredeemable gloom hung over and pervaded all' as if the slimy poison of depression had taken up residence. The miasmic atmosphere both outside and inside the House of Usher combine to produce and are produced by what David Miller calls 'the miasmic atmosphere of The Fall of the House of Usher' (Miller 1989, p. 195). Yet 'the rank miasma of the tarn' do not so much produce physical debility in accordance with the then prevailing miasmatic theory of disease as provoke mental derangement in Roderick Usher and terror giving way to horror in the first-person narrator/character after an initial shudder at the inverted image of the decaying trees reflected in 'the deep and dank tarn'.

The narrator tries to 'alleviate the melancholy' of Usher but perceives 'the futility of all attempt at cheering a mind from which darkness, as if an inherent positive quality, poured forth upon all objects of the moral and physical universe in one unceasing radiation of gloom'. This darkness is produced by the black sun of melancholy clouded over by the 'pestilent and mystic vapour' rising from 'the dim tarn' into which all the walls and turrets of the house looked down. The house as a whole is reflected in the tarn like its fungi covered walls and the decaying trees around it.

As a result the house participates in the same decay as the walls and in the same inversion as the trees, a house turned upside down in the narcissistic, reflecting and still waters of the tarn. If the house had been located next to a flowing stream, or even a canal as in one of Constable's paintings, it would not have been possible for things to have gone so awry. The tarn is made to blame for the mental ills that beset Roderick Usher. What the narrator calls 'the *physique*' of the House of Usher and its tarn produced a devastating effect on 'the *morale*' of Roderick Usher. Like Burton the anatomist of melancholy, the narrator of *The Fall of the House of Usher* sees some sort of sympathy or correspondence between the external physical environment and the internal metaphysical landscape.

Poe could be aptly renamed Poe(t) of the s(ub)lime and melancholy since in his writings in general and in his poems in particular he enunciates similar sentiments to those in *The Fall of the House of Usher*. In 'Dream-Land' he refers to 'Lakes that endlessly outspread/Their lone waters – lone and dead, – /Their still waters – still and chilly.' The way to this dream wetland is:

... by the swamp
Where the toad and the newt encamp, –
By the dismal tarns and pools
Where dwell the Ghouls, –
By each spot the most unholy –
In each nook most melancholy, –
There the traveller meets aghast
Sheeted memories of the Past –
Shrouded forms that start and sigh ...

In the swamp, Poe, like Strother, meets ghouls and ghosts. Similarly in 'The Lake – To –' he describes how as a youth he haunted 'a wild lake' whose 'loneliness' was 'lovely'. At night he would suddenly wake up, literally and/or metaphorically, to 'the terror of the lone lake' though 'that terror was not fright,/But a tremulous delight'. And although 'death was in that poisonous wave,/And in its gulf a fitting grave,' the young Poe(t) was able to make 'An Eden of that dim lake' perhaps by re-inverting the image of the upside down fallen world around the still waters of the lake into an image of Eden.

A number of psychoanalytic readings have been undertaken of Poe's work, most famously by Marie Bonaparte and more recently by Gaston Bachelard. For the former, 'that lake whose menacing waters glimmer so lugubriously through Poe's work' is 'a symbol of the dead Mother . . . the spectral Mother' (Bonaparte 1949, p. 146). Thus the ghoul which haunts the tarn is the lost Mother. And even the waters of the tarn are, according to Bachelard, 'the substance symbolic of death' (Durand 1969, p. 104). Gilbert Durand comments that 'the water which oozes out is a bitter invitation to the voyage without return'. A euphemism for death, the voyage without return is quite unlike the successful epic hero's return home following his descent into the underworld of the swamp like Martin Chuzzlewit's journey to Eden or Strother's journey into the Great Dismal Swamp. Durand goes on to argue that water's 'becoming is charged with terror, it is even the expression of terror', though I would argue that it is charged with horror as it is associated with the mother's body, with the becoming of life in the waters of the womb (Durand 1969, p. 104).

Poe's fascination with water, whether it be the waters of rivers, lakes, swamps, tarns or the sea, is indicative for Bachelard of a fascination with the 'waters' or milk of the mother's breast:

from a psychoanalytic point of view . . . all water is a kind of milk. More precisely, every joyful drink is mother's milk . . . Milk [is] . . . the first substantive in the order of liquid realities; more exactly, it is the first substantive known to the mouth . . . none of the values associated with the mouth is repressed. The mouth, the lips – here is where we know our first positive and specific happiness, the place where sensuality is permitted. (Bachelard 1983, pp. 117–18)

Yet for Poe, as he lost the waters of milk, every melancholic drink from the waters of the tarn in the dream-wetland is a bitter cup. Some waters are more milk than others and some drinks from them melancholic, others joyful.

Wetlands are generally the environmental waters of milk. Erich Neumann refers to 'the earth water as the milk of the earth body'

(Neumann 1955, pp. 47, 48). Wetlands are the moist womb and their living waters the breast of the Great Mother. Wetlands and living waters, womb and breast, are 'organs of receptivity and bounty' to use Melanie Klein's terms (Klein 1986, p. 74). Wetlands as womb are the source of life and wetlands as living waters are the first source of nourishment, the first object of love and the first object to be lost in modernisation, colonisation, drainage and so-called progress. The living waters of wetlands thus enact a sense of the loss of the breast as a loved object more than others. Yet these are the waters, in Melanie Klein's terms, of the good breast – white waters – whereas the water of wetlands have been construed traditionally in patriarchal western culture as the water of the bad breast – black waters – the milk of the dangerous *femme fatale* and entrapping spider woman. The milk or white water of the good breast, of the wetland as good breast, gives what Michelet called in relation to the sea 'refuge, warmth and repose', whereas the milk or black water of the bad breast, the wetland as bad breast, gives homelessness, coldness and uncanniness (Bachelard 1983, p. 119). Waters have been split by patriarchy.

For Freud melancholia (but why no Freudian theory of the phlegmatic?) is seen as 'regression from object cathexis to the still narcissistic oral phase of the libido' (Freud 1984, p. 259). This phase is narcissistic as all those references to images in water in Strother and Poe attest. Instead of investing desire in the object of love such as the breast and/or the mother and gaining some return of pleasure on that investment (the economic metaphor is appropriate), the melancholic 'wants to incorporate this object into itself, and in accordance with the oral or cannibalistic phase of libidinal development in which it is, it wants to do so by devouring it' (ibid. p. 258). In environmental terms, the melancholic wants to incorporate the nourishing qualities of the living waters of the wetland breast into himself by devouring it through drainage or filling, and even by creating artificial wetlands. The object of desire was initially an object of love which was later lost. For Freud melancholia, like mourning, 'may be the reaction to the loss of a loved object' (ibid. p. 253). The loved object which is lost for the melancholic is the breast of the mother and the water which is specifically breast milk is the water of the wetland, the first water which nourished life (on earth), the breast of mother earth.

Freud goes on to distinguish between mourning and melancholia by arguing that 'in mourning it is the world which has becomes poor and empty; in melancholia it is the ego itself' (ibid. p. 254). In mourning the world is experienced as loss whereas in melancholia the ego is experienced as lost. Freud outlined the process whereby melancholia establishes an 'identification of the ego with the abandoned object' (ibid. p. 258): 'the shadow of the object fell upon the ego ... as though it were an object, the forsaken object. In this way an object-loss was transferred into an ego-loss' (ibid. p. 258). As Mikkel Borch-Jacobson concludes in his discussion of melancholy, 'the ego becomes the object,' or even more succinctly, 'the lost object [is] me' as Nicolas Abraham and Maria Torok put it, as a way of denying or disavowing the loss of the loved object in narcissistic inversion (Borch-Jacobson 1989, p. 183; Abraham and Torok 1984, pp. 3–18).

Instead of seeing the object (the breast, the wetland) as lost, the melancholic ego sees itself as lost in a massive act of narcissistic disavowal and egotism. As a result the ego desires itself, or in the terms of Deleuze and Guattari 'it is the subject ... that is missing in desire, or desire that lacks a fixed subject; there is no fixed subject unless there is repression' (Deleuze and Guattari 1977, p. 26). Repression is to subjectivity as drainage is to wetlands; repression (and drainage) fix the flows of embodied subjects (and wetlands). Repression is constitutive precisely of melancholic subjectivity; subjectivity is melancholic and mournful. The subject desires itself as a product of a melancholic loss of the loved object of the mother and mother earth, of the mother's breasts and mother earth's breasts, the living waters of wetlands.

As wetlands are increasingly lost from the world and are lost as an object of love which nourishes life, both mourning and melancholia are experienced and exercised in relation to them. The world should mourn its wetlands which gave it life and nourished it, regarding the world as empty or in Gerard Manley Hopkins' apt word 'bereft' of them. Instead, it experiences this loss as a melancholic loss of its own ego, its own selfhood and sense of identity (Hopkins 1953, p. 51). Selfhood and identity are constituted in relation to something outside oneself against which one, the unitary discrete ego, stands in contradistinction. When that object is lost, as wetlands are, that loss is experienced in melancholia as a loss of the subject. The world is becoming empty of wetlands; the world is losing its wetlands. In patriarchy and filiarchy wetlands are the bad breast of the world but in these psycho-ecological terms they are the womb and breast of the world, the organs of receptivity and bounty as Melanie Klein puts it – in short, living waters.[6]

NOTES

1. Other examples of melancholic marshes in adolescent literature are the Swamps of Sadness in Michael Ende's *The Neverending Story* and the Dead Marshes in J. R. R. Tolkien's *The Lord of the Rings*. I am grateful to Christina Lupton for reminding me about the latter.
2. I am grateful to Stephan Millett for his comments on an earlier version of this section and for drawing my attention to Ross' discussion.
3. G. E. R. Lloyd (1970, p. 107) points out that 'the Greek terms *hygron* and *xeron* are wider than 'wet' and 'dry' in English, for *hygron* refers to both liquids and gases [as the elements of water and air show], and *xeron* especially, but not exclusively, to solids [as the element of fire shows]'. I am grateful to Stephan Millett for drawing my attention to this discussion.
4. I am grateful to Ann McGuire for drawing my attention to the article.
5. I am grateful to Margaret MacIntyre for reminding me about this story and its mire.
6. See Giblett 1994, pp. 55–62 for a discussion of Klein's work on the good and bad breast.

8

Marsh Monsters and Swamp Serpents: Horror of Horrors

In patriarchal western culture wetlands have not only been seen traditionally as places of disease and depression, but also as places of horror, often regarded as home to a horrific marsh monster or swamp serpent lurking in their murky, watery depths. Like the festering corpse beneath the mire, 'the serpent of the swamps ... lurks beneath the slough' (Soyinka 1973, p. 184). In the Slough of Despond lies coiled the monstrous serpent of the swamps which gives rise not only to marsh melancholy but also to downright horror. As the marsh or swamp itself is often represented as a place of horror, the swamp serpent or marsh monster is doubly horrific, the horror of horrors.

In many narratives, Bachelard argues, 'accursed places have at their centre a lake of shadows and horror' (Bachelard 1983, p. 101), or more precisely a swamp, marsh or fen, in short a wetland of black water in which there lives a monstrous and horrific 'thing'. This association between wetlands and the monstrous is a crucial one for bearing out the difference between the wetland conceived exclusively as a place of darkness and death as in patriarchal and filiarchal western cultures with its black waters and horrific monsters, or seen ambivalently as a place of both life and light, darkness and death with its living black waters, as in traditional or indigenous societies with their 'Snake Goddess' and 'Rainbow Serpent'. In this chapter I trace, deconstruct and decolonise this difference through a discussion of a variety of marsh monsters and swamp serpents including the Stymphalian birds of Greek mythology, Grendel in *Beowulf*, and Satan in Milton's *Paradise Lost*. In conclusion I compare and contrast the

179

patriarchal western view of marsh monsters and swamp serpents, with the Snake Goddess of Old Europe and with the Australian Aboriginal 'Rainbow Serpent' both associated with the living black waters of wetlands.

GRIM AND GREEDY CREATURES OF SWAMPS AND MARSHES

The sixth labour of the heroic Heracles (Hercules) in Greek mythology is to kill the Stymphalian birds, 'the countless brazen-beaked, brazen-clawed, brazen-winged, man-eating birds, sacred to Ares' which 'had flocked to the Stymphalian Marsh' (Graves 1955, p. 119). They are waterbirds and war-birds – wa(te)rbirds for short. From the marsh the wa(te)rbirds occasionally took to the air in 'great flocks to kill men and beasts by discharging a shower of brazen feathers and at the same time muting a poisonous excrement which blighted the crops' (ibid. p. 119). No doubt this poisonous excrement is produced by the birds because they live in and feed greedily on the ostensibly decaying and rotting life of the marsh. In short, they are coprophagic. They are later described by Lucretius in the first century BCE as 'the foul birds that haunted the Stymphalian lake' (Lucretius 1994, p. 129). They are waterfoul, not waterfowl.

The birds represent the antithesis of, or perhaps more precisely an offence (in the sense of attack, annoyance and transgression) towards, agriculture. In response, even resistance to this myth, my reading is a defence of the marsh and its birds. The marsh and the martial are allied in the myth against agriculture and men (andros). The ostensibly martial marsh birds are figured as mis-andropic in this foundational myth of patriarchal western culture which displaces the blame onto the victim for the war patriarchal men have fought against the marsh from, no doubt, before heroic Heracles' time. This martial anti-marsh campaign culminates in fascist Mussolini's boast that the war against the Pontine Marshes was the kind of war he liked to fight, and win.

Instead of men and agriculture being shown as antagonistic to the marsh, the marsh is made out to be monstrously martial towards men and agriculture. The martial, men and agriculture (and the matrix) have been allied against the marsh, the maternal and marshbirds in western culture. Marsh and agriculture are opposed but they can (or should) co-exist whereas the Stymphalian birds are made to represent a martial/marsh-al oral and anal sadism which would destroy agricultural crops and vanquish the marsh over agriculture. The oral and anal sadism which men have exercised against the marsh by eating it up, grazing their animals on it (consuming its resources), administering an enema to it (draining it) and defecating in it ('sanitary land-filling' it) is here displaced onto anthropophagic, or more precisely androphagic, coprophagic and coprophilic waterbirds. As they feed on shit and shit shit, their shit is not manure but poison.

Heracles is initially unable to drive the birds away with his arrows. Typically, even archetypically, he finds that 'the marsh seemed neither solid enough to support a man walking, nor liquid enough for the use of a boat' (Graves 1955, p. 119). The wetland for Heracles, and for hundreds of generations of heroic westerners since him, is neither solid land nor deep water but mediates (between) the two. The wetland is centrally construed in patriarchal and filiarchal western culture as a problem of transportation. So it is for Allnutt in *The African Queen*, and is especially for the hero on a mission or quest or with a job to do which requires him to traverse the place with some goal in view or mind. Yet rather than a hero on a single-minded mission to conquer the marsh by crossing it (though in one version of the myth Heracles drained the marsh (Graves 1955, p. 120)), Heracles' heroic labour is to kill the horrific wa(te)rfoul.

Heracles 'labour' performed by a birthing metaphor 'glorifies' or gives birth to himself, according to Robert Graves, as

the healer who expels fever demons, identified with marsh-birds. The helmeted birds shown on Stymphalian coins are spoon-bills, cousins to the cranes which appear in English medieval carvings as sucking the breath of sick men. They are, in fact, bird-legged Sirens, personifications of fever . . . The spoon-bill is closely related to the ibis, another marsh-bird, sacred to the god Thoth, inventor of writing. (ibid. pp. 120–1)

The Stymphalian birds personify, or more precisely 'femininify' as they were Sirens and women according to some accounts (ibid. p. 120), the miasmatic diseases to which the marsh gives rise. How spoonbills, a benign species of waterbird which feed on crustaceans and small fish by sweeping its spoon-shaped bill from side to side under the water, came to femininify disease and to be blamed for the war men have fought against the marsh is not only a mystery but also a patent injustice (Kingsford 1991, p. 70).

A limited explanation of both is offered by Bachelard when he suggests that the marsh in this myth becomes what he calls '"stymphalized". It becomes the black swamp where live monstrous birds, the Stymphalides, "nurslings of Ares, which shoot their feathers like arrows, which ravage and soil the fruits of the earth, which feed on human flesh". This stymphalization . . . corresponds to a particular feature of melancholy imagination' (Bachelard 1983, p. 101). I suppose only a deeply melancholic, even manic depressive, imagination could see a spoonbill as a bird which wars on humans, destroys their crops by excreting on them and eats their flesh, as, in short, both anally and orally sadistic. As a result of such a diet it is hardly surprising then that they 'mute' poisonous excrement. Such excrement is thus the result of their diet of human crops and flesh rather than the product of the marsh in which they live.

Marsh monsters are the product not only of a melancholic imagination but also, and in turn, of a miasmatic theory of monstrosity. Not only does the marsh 'cause' physical and 'mental' illness due to the vapours which rise from it, but also the miasma give birth to monsters. In *Beowulf* Hrothgar describes Grendel the monster along with his mother in miasmatic terms as two:

huge wanderers of the marches guarding the moors, alien spirits . . . They dwell in a land unknown, wolf-haunted slopes, wind-swept headlands, perilous marsh-paths, where the mountain stream goes down under the mists of the cliff, – a flood under the earth. It is not far hence, in miles, that the lake stands over which hang groves covered with frost: the wood, firm-rooted, over-shadows the water. There may be seen each night a fearful wonder, – fire on the flood! . . . That is no pleasant spot. Thence rises up the surging waters darkly to the clouds, when the wind stirs up baleful storms, until the air grows misty, the heavens weep (*Beowulf* 1950, pp. 89, 90),

as they do also in *Sir Gawain and the Green Knight* where 'the heavens are lifted high, but under them evilly/ mist hangs on the moor, melts on the mountain' (*Sir Gawain* 1975, p. 77).

The wetland not only mixes the elements of water and earth, and air and water when the atmosphere is misty in the temperate zone (and dripping wet in the tropics), but also fire and water when marsh gases are ignited in the temperate zone. The traditional dragon breathing fire is arguably just a marsh (monster) 'breathing' ignited marsh gases. Fire is also associated with hell, with the burning fires of brimstone, so the wetland in patriarchal western culture is a kind of watery hell.

Grendel and his mother are alien spirits because they are 'Cain's kindred' (*Beowulf* 1950, p. 25) condemned by God to the wetland hell guilty, presumably like Cain, of fratricide though who his or her brother is, is not specified. The worst of the two is 'that grim spirit . . . called Grendel, the renowned traverser of the marches, who held the moors, the fen and fastness; unblessed creature, he dwelt for a while in the lair of monsters, after the Creator had condemned them' (ibid. p. 25). Grendel, however, has not only been condemned but also been made into the personification, or more precisely 'monstrification', of damnation. He is described as 'the grim and greedy creature of damnation, fierce and furious', and as 'the demon, the dark death-shadow' who 'in the endless night . . . held the misty moors' (ibid. pp. 26 and 28). Grendel is a demonic agent, or 'angel', of death who lives in the wetland realm of darkness.

Grendel's mother is not much better. She is 'monstrous among womenkind . . . who must needs inhabit the dread waters, chilling stream' (ibid. p. 84). Grendel's mother lives in the dreadful aquifer, the flood beneath the earth, which is even worse than the flood on or over the earth. Grendel's mother is an instance of what Barbara Creed calls 'the monstrous-feminine' (Creed 1986, pp. 44–70), or more precisely the monstrous-maternal, a patriarchal troping of the mother as monster. Grendel and his mother are both specifically marsh monsters and a part of a long genealogy.

Grendel is no doubt a prototype for Satan in Milton's *Paradise Lost*. Milton's Satan is a swamp serpent and marsh monster, or more precisely a miasmatic marsh monster. He is described as 'involved in rising mist' (IX, 75) and as 'wrapped in mist/Of midnight vapour' who glides 'obscure' (IX, 158–9) 'like a black mist low creeping' (IX, 180). Satan is a miasmatic marsh monster because as a fallen angel he has been 'mixed with bestial slime,/

This essence to incarnate and imbrute' (IX, 165–6). Satan personifies, or more precisely monstrifies, slime and wears as his proper wardrobe the miasma which arise from it. He monstrifies slime because he was 'engendered in the Pythian vale on slime' (X, 530). He is born from slime and made of slime; he is slime by birth and slime by nature.

Slime, as we saw in Chapter 2, has been associated with the excremental, especially when faeces are not hard and dry nor soft and wet but in-between both sets of categories. For Milton it seems, as Stephen Greenblatt argues for Martin Luther, 'the Devil dwells in excrement' (Greenblatt 1982, p. 12). As a slimy swamp serpent his mode of propulsion is not restricted to any one kind, but varies according to the (wet)landscape over which he is travelling. As he does not stick to one mode of propulsion in one element he is aberrant: 'so eagerly the Fiend/ O'er bog or steep, through strait, rough, dense, or rare,/With head, hands, wings, or feet pursues his way,/And swims or sinks, or wades, or creeps, or flies' (*Paradise Lost* II, 947–50). Satan is even a flying miasmatic marsh monster and swamp serpent. He is monstrous because he combines the physical characteristics of birds (he flies in the air like the Stymphalian birds), of fish (he swims in water), reptiles (he creeps on land), and quadrupeds or bipeds (he sinks or wades in slime and swamp).

Satan, in a word, swarms and hence comes under the Biblical interdiction that 'every swarming thing . . . is an abomination' (Leviticus 11:41). Swarming creatures, Mary Douglas has commented, are:

both those which teem in the waters and those which swarm on the ground. Whether we call it teeming, trailing, creeping or swarming, it is an interminable form of movement . . . 'swarming' which is not a mode of propulsion proper to any particular element, cuts across the basic classification. Swarming things are neither fish, flesh nor fowl. Eels and worms inhabit water, though not as fish; reptiles go on dry land, though not as quadrupeds; some insects fly, though not as birds. There is no order in them . . . As fish belong in the sea so worms belong in the realm of the grave with death and chaos. (Douglas 1966, p. 56)[1]

Milton's Satan is a worm which belongs in the tomb of the swamp.

By swarming in the elements of earth, air, fire and water Satan is an abominable monstrosity who upsets the order of things. The late nineteenth-century Australian writer 'Tom Collins' (pseudonym of Joseph Furphy) maintained in *Such is Life* that: 'Nature, by a kind of Monroe Doctrine, has allotted the dry land to man [*sic*] and various other animals; the water to fish, leeches, etc.; the air to birds, bats, flies, etc.; the fire to salamanders, imps, unbaptised babies, etc.; and she strictly penalises the trespass of each class on the domain of the other'. Milton's Satan is severely damned because he trespasses on all domains, in all four elements.

Nature, according to Collins by a kind of isolationist/separatist American/European hemispherical divide, has not allotted the wetland to 'man'. 'Man' belongs in the dryland hemisphere on the earth's surface according to Collins, not in the wetland hemisphere in the earth's depths. But the wetland does not belong to any one element as we have seen. The swamp, like Satan, trespasses on every domain. The wetland is a vital part of the whole biosphere, not just a hemisphere.

Milton not only seems to have drawn on Grendel for his descriptions of Satan's physical features but also seems to hark back to a figure of the pre-Socratic cosmogony ironically called Heracles. In the pre-Socratic cosmogony, at least in christian Athenagoras' second century CE account 'water was the origin for the totality of things . . . and from water slime was established, and from both of them was generated a living creature, a snake with a lion's head growing on to it' (Kirk et al 1983, p. 25),[2] the king of anti-beasts, or monsters, and a kind of monstrous antithesis to or parody of the Sphinx with its human head and lion's body. The Sphinx, Barbara Creed has suggested, 'is an ambiguous figure; she knows the secret of life and is thereby linked to the mother-goddess but her name, which is derived from 'sphincter,' suggests she is the mother of toilet training, the pre-Oedipal mother who must be repudiated by the son so that he can take his proper place in the symbolic' (Creed 1986, p. 61). The riddle of the Sphinx could therefore be more aptly called the riddle of the sphincter.

As there is some sort of homology between the body and the (wet)landscape, the riddle of the Sphinx might even be seen as the riddle of the sluice gate. The Sphinx could be seen as the mother

of swamp-drainage whereas Satan is the monster of the swamp
who is linked to the pre-'pre-Oedipal' swamp mother. As Luther's
Satan dwells in excrement, as Milton's Satan was engendered in
slime, and as Dante's Satan's anus is the centre of Hell, these
monstrous swamp serpents parody and solve the riddle of the
sphincter. These Satans represent and act out a fixation with the
anal and excremental as if their creator's toilet-training was too
successful in that, by repudiating the mother, their creators
necessarily became anally fixated; by repudiating the traditional
nurturer, the giver (the subject and agent) of softness and warmth,
they became fixated on objects, on cold, hard things, and horrified
(and fascinated) by abjects, to use Julia Kristeva's term, by slime
and swamps. To repudiate that repudiation would be to retrieve
the mother-goddess of our pre-history and to affirm slime and
swamps as a necessary part of life. Dante, Luther and Milton all
seem to be playing out similar anxieties associated with birth,
excrement and the mother.

THE SWAMP AS MONSTER

As an increasingly rationalistic and secularised culture, especially
its urban intellectuals disconnected from their bodies, no longer
believed in monsters, the marsh and swamp came to be seen less
as the dwelling place of monsters, (though this association still
persists especially in popular culture) and more as a monstrous
place in itself, or at least of monstrous floristic or 'vegetable'
growths. The attitude of American Victorianism to the swamp as
monstrous is exemplified in William Gilmore Simms' poem 'The
Edge of the Swamp' where:

> . . . A rank growth
> Spreads venomously round, with power to taint;
> And blistering dews await the thoughtless hand
> That rudely parts the thicket Cypresses,
> Each a ghastly giant, eld and gray,
> Stride o'er the dusk, dank tract, – with buttresses
> Spread round, apart, not seeming to sustain,
> Yet linked by secret twines, that underneath,
> Blend with each arching trunk. Fantastic vines,
> That swing like monstrous serpents in the sun,
> Bind top to top, until the encircling trees

Group all in close embrace. Vast skeletons
Of forests, that have perish'd ages gone,
Moulder, in mighty masses on the plain;
Now buried in some dark and mystic tarn . . .
The place, so like the gloomiest realm of death,
Is yet the abode of thousand forms of life, –
The terrible, the beautiful, the strange, –
Wingéd and creeping creatures, such as make
The instinctive flesh with apprehension crawl . . .
                              (Simms 1853, pp. 201–2)

The swamp is the region of death which gives birth to monstrous life. It is hardly surprising then that the place makes the flesh crawl. After all it is 'the slimy green abode' (ibid. p. 202) of the crocodile, 'the mammoth lizard', in which and with whom for Simms a modern-day Aesopian fable is played out with a butterfly.

In the swamp, or at least on its edge, a butterfly is 'a wanderer in a foreign land' who 'lights on the monster's brow'. But the crocodile, 'the surly mute/Straightway goes down' (ibid. p. 202) yet so suddenly that:

The dandy of the summer flowers and woods,
Dips his light wings, and soils his golden coat,
With the rank waters of the turbid lake.
Wondering and vexed, the pluméd citizen
Flies with an eager terror to the banks,
Seeking more genial natures, – but in vain.
Here are no gardens such as he desires,
No innocent flowers of beauty, no delights
Of sweetness free from taint. The genial growth
He loves, finds here no harbour. Fetid shrubs,
That scent the gloomy atmosphere, offend
His pure patrician fancies. On the trees,
That look like felon spectres, he beholds
No blossoming beauties; and for smiling heavens,
That flutter his wings with breezes of pure balm,
He nothing sees but sadness – aspects dread,
That gather frowning, cloud and fiend in one,
As if in combat, fiercely to defend
Their empire from the intrusive wing and beam.
                              (ibid. p. 203)

The edge of the swamp is not the place of smiling heaven but scowling hell. It is the dark underside of the garden of Eden, what David Miller has called dark Eden (Miller 1989). Simms concludes his poem by drawing the moral of the fable that 'the example of the butterfly be ours' in that we should flee these 'drear borders' for 'a more genial home', for a homely, canny place, rather than what 'these drear borders offer us to-night' (ibid. p. 203) to stay in the uncongenial, unhomely and uncanny place of the swamp.

Simms' poem has been seen by Louis Rubin as 'an allegory of the writer in the South' (Rubin 1989, p. 52) of the United States with perhaps the writer as butterfly who needs to escape the gaping and ravenous maw of the allegorical crocodile and swamp of the deep south which would drag the writer down into its depths and consume him, or her. It may, however, be more appropriate to read the moral fable and the aesthetic allegory within a larger framework as a manifestation and playing out of certain anxieties, repressions and sublimations about southern American culture, the swamp and sexuality. Southern writers for Rubin seem to need to deny and repress their roots in the slimy swamps, Everglades and Bayous of the deep south in preference for the sublimated bright city lights of the north.

Similar sentiments are expressed by Harriet Beecher Stowe in her novel *Dred* first published only three years after Simms' poem. It also sees 'those desolate regions' of the Great Dismal Swamp as monstrous in themselves:

it is said that trees often, from the singularly unnatural and wildly stimulating properties of the slimy depths from which they spring, assume a goblin growth, entirely different from their normal habit. All sorts of vegetable monsters stretch their weird, fantastic forms among its shadows. There is no principle so awful through all nature as the principle of *growth*. It is a mysterious and dread condition of existence, which, place it under what impediment or disadvantage you will, is constantly forcing on; and when unnatural pressure hinders it, develops in forms portentous and astonishing. The wild, dreary belt of swamp-land which girds in those states scathed by the fires of despotism is an apt emblem, in its rampant and we might say delirious exuberance of vegetation, of that darkly struggling, wildly vegetating swamp of human souls, cut off, like it, from the usages and improvements of cultivated life. (Stowe 1856, p. 274)

Stowe uses the swamp as a figure for southern slavery, but also contrasts the wilderness of southern swamps with the cultivation of cities, agriculture and the north. The swamp for Stowe is horrific not because it is a region of death and decay, but because it is one of exuberant and monstrous life.

As a wetlands' scientist has put it 'wetlands . . . are among the most productive of all natural habitats' (Niering 1966, p. 166). Rather than a source of horror, the fecundity and fertility of wetlands should be a reason for their conservation and a cause for their celebration, not the motivation for their denigration and destruction as monstrous places, or as unhomely homes for monsters. This revaluation of the swamp could even give rise, as it does in Stowe's novel, to the kind of 'discourse [which] was like the tropical swamp, bursting out with a lush abundance of every kind of growth – grave, gay, grotesque, solemn, fanciful, and even coarse caricature, provoking the broadest laughter' (Stowe 1856, pp. 305, 306). This would be a counter-discourse to the standard swampspeak of horror and monstrosity.

The physical and floral features of the swamp, however, are highly variable season by season as Dorothy Langley's novel *Swamp Angel* set in the swamps of southeast Missouri around the settlement of Weary Water shows. The seasons of the swamp, a swamp phenology, can move through the frozen waters and mystery of dormant life in winter, through the thawing of snow and ice and the celebration of new life in spring, through the stinking horror of dying life and stagnant waters in summer to the unmentionable death, decay and black waters of autumn. In addition, this novel not only subscribes to a miasmatic theory of illness as we saw in a previous chapter, but also to the patriarchal theory of the swamp which feminises it as passive, especially in summer:

the swamp had known a brief period of loveliness in the spring when trilliums and hepaticas covered the raw ground. It would be beautiful again in winter, with snow setting out the gnarled trees and crippled fences in patterns as wild, as mysterious with secret life, as a landscape painted in an opiate dread. But now, under the terrible down-bearing heat of late summer, it lay inert and panting in the sun, its hard rutted roads choked with yellow dust, its trees baked saffron and standing motionless. (Langley 1982, p. 31)

The swamp changes seasonally between the new life of spring, the secret life of winter and the dying life of summer. The death of autumn is not even mentioned, and is probably unmentionable.

Part of the horror of the swamp is its mutability. It is not fixed and static, but moving and dynamic. It does not keep still like a compliant woman for patriarchal agriculture to plough and inseminate. It is unsuitable for agriculture, but in a rerun of the myth of the Stymphalian birds it is represented as intractable and recalcitrant towards agriculture, especially in summer when the sun beats down on it. The sadistic sun is gendered as masculine which masters, or tries to master, the supine and panting feminine body of the swamp.[3] The sun and the swamp merely play out in correspondence the gender relations between the men and women of the settlement of Weary Water where:

> flat-breasted women moved sluggishly about their household tasks or toiled up the slimy banks of the river carrying heavy wooden pails of silt-clouded water, their eyes as blank and incurious as the eyes of a beast of burden . . . [while] men tilled their grudging fields with rusty implements, sweating and swearing like so many prisoners digging with their finger nails at a dungeon floor. (Langley 1982, p. 31)

The flat-breasted and sluggish women are analogous to the flat-breasted and slug-ish swamp. The flat-breasted swamp is not the rain-giving and rounded breasts of the enclosed slopes, and hardly much better than the shrivelled and dried-up dugs of the desert. The ambivalent flat breast of the swamp is situated, or mediates, between the good breast of the rain-giving slopes, and the bad breast of the desert.

THE DREDGER AS MONSTER

Yet in Graham Swift's postmodernist novel *Waterland* set in 'the devil-ridden Fens' which traditionally 'yield so readily to the imaginary – and the supernatural' (Swift 1984, pp. 15, 72) the only monster is an orally sadistic and satanic dredger. This product of modern industrial technology is described as 'a mud-sucker,' 'a sludge-extractor' and as 'a monstrous deformity' (ibid. p. 299) with its 'giant snout of the bucket-ladder [which] is biting, gnawing with its rotating teeth into the soft, defenceless belly of the river-bed' (ibid. p. 308) which ends up in 'the bowels of the dredger' (ibid. p. 302). Instead of the spoonbills as in the ancient Greek myth of the Stymphalian waterbirds eating men, in this postmodern inversion, even subversion, of the myth the man-made dredger eats the wetland. The spoonbill of the Stymphalian birds becomes the giant snout of the monstrous dredger; the brazen beaks and claws of the birds are transformed into the gnawing and rotating teeth of the bucket-ladder; the neither solid nor liquid marsh is reworked into the soft belly of the fen river-bed; and the offensive and martial marsh pacified into the defenceless and maternal fen.

A similar monstrous dredger can be found in Liam Davison's *Soundings*, a kind of Australian *Waterland*. In Davison's novel an orally sadistic German steam dredger 'anxious to sink its hardened steel teeth into the [wet]land . . . brought the brute force of the war with it' to bear on the drainage of the swamps (Davison 1993, pp. 97, 99, 103, 172). The war against the wetlands, the kind of war which Mussolini liked to fight against the Pontine marshes, and win, and the kind of war which Heracles fought against the marsh, is carried out by monstrous war machines. Similarly in *Waterland*, 'the essential desolation of the Fens . . . the untamed swamps' (Swift 1984, p. 137), the wet wilderness and the wild wetlands, were desolated (the desolation desolated) and tamed (the untamed tamed) by the orally sadistic monsters of modern industrial technology.

Swift and Davison invert nicely, even subvert radically, the traditional association between the monstrous and the wetland, and the modern antithesis between industrial technology and the wetland, by having not wetlands as monstrous but modern, industrial technology. The inescapable conclusion is that the war against the wetlands by dredging, draining and other types of wetlands engineering was fought by man-made monsters. Instead of the marsh being monstrous, it is modern technology; instead of the marsh being home to unhomely monsters, the antithesis of the city, it is the city with its modern technology which is home to monstrous dredgers.

The monstrous dredgers of these two novels can be read analogously with a shift that was taking place at about the same time in Hollywood films and more broadly in American culture around environmental concerns. Monster figures, as Michael Ryan and Douglas Kellner have argued, 'can be used to affirm the existing order in that they represent threats to normality which are purged'. In what they call 'classical monster films a reassuring social order is restored through the successful operations of conservative institutions and authority figures' (Ryan and Kellner 1990, p. 179). No doubt Grendel and his mother in *Beowulf* and the Satan of *Paradise Lost* serve to some extent this conservative function. But, as Ryan and Kellner go on to argue, 'monsters can be used critically and deconstructively if they draw attention to particularly monstrous aspects of normal society ... In most contemporary monster films no reassuring vision of restored order is affirmed' (ibid). Monsters are double or split figures which have a critical and deconstructive as well as conservative and institutionalising function.

The monstrously pathological can be used to show up the monstrously normal; the monstrously pathological dredging war machines can be used to show up the monstrously normal war against the wetlands; the monstrous dredgers of *Soundings* and *Waterland* draw attention to the monstrously normal practice of draining wetlands. Even Grendel, his mother and Satan given their attraction as 'baddies' over their 'goody' counterparts, show up the repressions, desires and fears in the cultures of their periods. Indeed, Ryan and Kellner go so far as to argue that 'all monster figures are immanently critical of reigning social norms, since what cultures project as 'monstrous' in relation to 'normality' is frequently a metaphor for unrestrained aggression and unrepressed sexuality of the sort upon whose exclusion the maintenance of civility depends' (ibid). Monsters are also symptomatic of the unrepressed slimy and swampy upon whose exclusion the city is dependent for its existence.

THE (SWAMP) THING

The symptomatology of swamps monsters involves their nomenclature as 'the Thing'. In the swamp of Victor Kelleher's *Brother Night* 'where the mist rises and the gnarled trees grow thick about the stagnant hollows' there was to be found, according to tradition, "a monstrous thing ..., a great twisted shape that had somehow escaped from our dreams' (Kelleher 1990, pp. 8, 9).[4] The monster in Peter Tremayne's *Swamp* is described by one character as 'that thing', as is the monster in *Danny Dunn and the Swamp Monster* and a whole comic series has been built around 'The Swamp Thing' (Tremayne 1985, p. 189; Williams and Abrashkin 1972, p. 101).[5] The 'Thing' has been defined by Julia Kristeva as: 'the real that does not lend itself to signification, the centre of attraction and repulsion, the seat of sexuality from which the object of desire will become separated ... the Thing is an imagined sun, bright and black at the same time' (Kristeva 1989, p. 13).

As the slimy real that does not lend itself to signification, the Swamp/Thing gives rise to the horror of swamps (and to swamps of horror) and to inarticulate sounds of horror; as a bright and black imagined sun, it eclipses the black sun of depression and melancholia. Although marsh miasma give rise ultimately to marsh monsters, the latter are a way of coping and dealing with the melancholy (and even mourning) associated with the loss of wetlands. And although the monster gives rise to inarticulate sounds of horror, the monster has still been named as such, as 'baleful' and so on.

The monster as bright and black sun and the naming of the monster function as a sublimation of the symptoms of repression. Freud argued that 'repression degrades a process of satisfaction to a symptom' and that 'symptoms are created in order to remove the ego from a situation of danger' (Freud 1979, pp. 246, 302; see also p. 284). The modern city as cultural symptom is created to remove the city itself from a situation of danger brought about by the repressions of the satisfying processes of the swamps it has degraded in order to establish itself. Sublimation merely extends and heightens this process (see Chapter 2, Figure 1).

Kristeva has shown how the two psychosomatic processes of symptom and sublimation are linked by arguing that

the symptom [is] a language that gives up, a structure within the body, a non-assimilable alien, a monster, a tumour, a cancer that the listening devices of the unconscious do not hear, for its strayed subject is huddled outside the paths of desire. Sublimation, on the contrary, is nothing else than the possibility of naming the pre-nominal, the pre-objectal, which are in fact only a trans-nominal, a trans-objectal. In the symptom, the abject permeates me, I became abject. Through sublimation I keep it under control. The abject is edged with the sublime. It is not the same moment on the journey, but the same subject and speech bring them into being. (Kristeva 1982, p. 11)

In other words, the abject or the slimy is the obverse of the sublime as we saw in Chapter 2. The city as symptom of the will to fill slimy wetlands inscribed on the surface of the earth produces the sublimated monsters of its dredgers, and of its dank and dark underside, its drained swamps and festering sewers. Yet this is intolerable for the city which must needs keep abjection and the uncanny under control, or risk being swamped by it, by naming the monstrous as such.

Herein for Kristeva lies the difference between 'the erotic Thing' and 'Object of desire' whereby the former is 'inscribed within us without memory' (Kristeva 1989, p. 14) or at least without individual memory, though not without 'race' or cultural memory, and the latter is inscribed within our individual memories and the official histories of our cities. Kristeva goes on to differentiate between the Thing on the one hand as 'the 'something' that, seen by the already constituted subject looking back, appears as the unspecified, the unseparated, the elusive, even in its determination of actual sexual matter [whereas on the other] Object [sic] [is] the space-time constant that is verified by a statement uttered by a subject in control of that statement' (ibid. p. 262, n.7). The monstrous thing goes back to the abject and uncanny which can then give rise to the object through the psycho(eco)logical/sexual process of sublimation and the aesthetics of the sublime.

The sublime, as Terry Eagleton has recently reminded us, 'for Marx as for Kant, is *Das Unform*: the formless or monstrous' (Eagleton 1990, p. 213). The swamp monster and the monstrous swamp have a shape (which are similar to each other, even homologous) but they do not have a Platonic form. They do not con-form to a pre-existing or transcendental pattern or ideal, but are variable and mutable. They are sublime and sublimated. Monsters perform the central function of art for Freud which is to sublimate repudiated or repressed elements of sexuality by producing socially acceptable daydreams (Freud 1985, pp. 140, 141). But they also show up the monstrously normal.

In the aesthetics of the sublime, Lyotard argues, 'art does not imitate nature, it creates a world apart . . . in which the monstrous and formless have their rights because they can be sublime' (Lyotard 1989, p. 202).[6] Perhaps it is hardly surprising then, as Eagleton notes, that 'Kant associates the sublime with the masculine and the military' (Eagleton 1990, p. 90), especially in the light of the war against wetlands. This association bears out further Zoë Sofoulis' contention discussed in Chapter 2 that slime is the obverse of the sublime for the masculine and military sublime sublimates, or wars, on the feminine and pacific slimy by rendering it monstrous thus trying to justify its own taming of the wet wilderness through drainage and wetland-filling.

Monsters not only serve this function in relation to sublimation, but also serve an important, and related, function in the taxonomies of natural history. Monsters, François Jacob argues, 'always bear resemblances, but they are distorted resemblances which no longer correspond to the normal action of nature' (Jacob 1973, p. 27).[7] In other words, monsters correspond to the pathological action of nature, but a nature pathologised by culture. Or as Michel Foucault puts it, monsters form 'the background noise, as it were, the endless murmur of nature' (Foucault 1970, p. 155). Monsters are the white noise or interference in the communication channel/canal through which human 'culture' talks to itself about non-human 'nature', including black waters, in that endless murmur of conversation of which nature documentaries are only an institutionalised part. Without that noise communication would not be possible. No doubt marsh monsters and swamp serpents serve these socially useful functions, but at the cost of psychogeopathologising the marsh and swamp themselves, at least within patriarchal western culture.

Historically within patriarchal culture the creation of monsters points to the ongoing struggle between two orders, the old order of traditional matrifocal swamps and the new order of patriarchal natural history. With the rise of 'the new biology' Jacob quotes Isidore Geoffrey Saint-Hilaire to the effect that 'monstrousness is no longer a random disorder, but another order, equally regular and equally subject to laws: it is the mixture of an old and a new order, the simultaneous presence of two states that ordinarily succeed one another' (Jacob 1973, p. 124). Jacob goes on to add that 'all types of monstrosity are not possible' and that 'monsters can be classified' in 'teratology', as 'nature' more generally is categorised in phenology and taxonomy (ibid. p. 124). The taxonomy of monsters is a way of mastering the monstrous. Monsters are produced by combining elements of known creatures, a head from there, a torso from elsewhere, a lion's head and a snake's body or a human head and a lion's body.

THE SNAKE GODDESS

The marsh monsters and swamp serpents of patriarchal times hark back to and are a vestige of what Marija Gimbutas calls the 'mistresses of waters', the bird and Snake Goddess of 'old Europe' (6500 BCE to 3500 BCE) who 'rules over the life-giving force of water' (Gimbutas 1982, p. 112). Although they appear as 'separate figures' according to Gimbutas, they are 'a single divinity. Their functions are so intimately related that their separate treatment is impossible. She is one and she is two, sometimes snake, sometimes bird. She is the goddess of waters and air, assuming the shape of a snake, a crane, a goose, a duck, a diving bird' (ibid. p. 112). And even a spoonbill if the Greek myth of the Stymphalian birds is anything to go by. In this pre-Indo-European tradition, the mixing of waters and air is the life-giving milk of the earth.

Moreover the wetland is also the womb and its living waters the breast of mother earth, or more precisely, of the Great Goddess. Gimbutas goes on to argue that:

in contrast to the Indo-Europeans, to whom Earth was the Great Mother, the Old Europeans created maternal images out of water and air divinities, the Snake and Water Goddess. A divinity who nurtures the world with moisture, giving rain, the divine food which metaphorically was also understood as mother's milk, naturally became a nurse or mother. (ibid. p. 142)

Grendel's mother is a patriarchal reworking of this figure as are the flat-breasted and sluggish swamp and women of Weary Water.

The double Snake and Water Goddess is the mother as nurse which is an aspect or development of the Great Goddess of life, death and regeneration. Whereas the Snake and Water Goddess nurtures life, the Great Goddess for Gimbutas was the 'giver of life and all that promotes fertility, and at the same time she was the wielder of the destructive powers of nature' (ibid. p. 152). Also unlike the Snake and Bird Goddess who is 'the feminine principle', the Great Goddess was 'not entirely feminine. She was androgynous' in whom 'the aspects of death and life are inextricably intertwined' (ibid. p. 196) as they are in wetlands, at least from a psycho(eco)logical perspective whereas from a patriarchal point of view they are a place of death, disease, excrement, melancholy and the monstrous which would inextricably entwine the unsuspecting male in their clutching and (s)mothering embrace.

In nineteenth-century American culture David Miller argues that 'this mingling [in the swamp's ecology] of forces normally thought to be opposed – life and death, good and evil – undercut the fundamental assumptions connected with the sentimentalist quest for purity and flight from death' (Miller 1989, p. 10). The confrontation with the swamp not only involved coming face to face with death, and even with one's own death (as Strother found, and repressed), but also with the new life which springs from death (as Stowe found, and stymphalized), to the horrific realisation that life goes on after one's own death, that life does not end when one ends.

The Great Goddess, Gimbutas goes on to argue further,

as a supreme Creator who creates from her own substance . . . is the primary goddess of the Old European pantheon. In this she contrasts with the Indo-European Earth-Mother, who is the impalpable sacred earth-spirit and is not in herself a creative principle; only through the interaction of the male sky-god does she become pregnant. (Gimbutas 1982, p. 196)

The Great Goddess brings life out of death herself. She enables and represents symbolically 'the cyclic returning and renewal of life' (ibid. p. 236). She represents that moment when death gives over into new life. The moment of new life, in turn, is represented by the Snake and Water Goddess. On the one hand:

the Snake and Bird Goddess create the world, charge it with energy, and nourish the earth and its creatures with the life-giving element conceived as water. The waters of heaven and earth are under their control. The Great Goddess [on the other] emerges miraculously out of death . . . and in her body new life begins. She is not the earth [unlike the Earth Mother of Indo-European

culture], but a female human, capable of transforming herself into many living
shapes. (ibid. p. 236)

The Great Goddess is fecund and protean like the wetland itself.

The 'Rainbow Serpent' and Living Black Waters

The 'Rainbow Serpent' of Australian Aboriginal cultures can be
likened to the Snake Goddess of Old Europe and contrasted with
the swamp serpents and marsh monsters of patriarchy. Yet the
'Rainbow Serpent' predates both as 'the appearance of the Rainbow Serpent belongs to the years between 9,000 and 7,000 years
ago which would make the Rainbow Serpent myth the longest
continuing religious belief documented in the world' (Flood 1995,
p. 171).[8] This 'myth' or 'religious belief' was first noted formally
by an anthropologist in 1926 when A. R. Radcliffe-Brown
announced to his fellow anthropologists that 'there is found in
widely separated parts of Australia a belief in a huge serpent
which lives in certain pools or water-holes. This serpent is
associated, and sometimes identified, with the rainbow'
(Radcliffe-Brown 1926, p. 19). This serpent could also be regarded
as a 'dreaded monster' who was 'at times tricky and malignant.
But he could do a good turn to men [sic] already possessed of
some magical quality' (ibid. p. 20).

In other parts of Australia, the serpent 'swallows human beings
whole' and is even in one place 'endowed not with huge teeth
only, but also with a special craving for the black-fellow' (ibid.
p. 21). The Serpent could be regarded as an orally sadistic
monster and so a bringer of death. But the Serpent is the protector
of water-holes and so of water with its life-giving properties. The
'Rainbow Serpent', Charles Mountford argued, 'is essentially the
element of water' (Mountford 1978, p. 23). Or, as Radcliffe-Brown
argued in a follow-up article on the subject, the Serpent even
'represents the element of water which is of such vital importance
to man [sic] in all parts of Australia' (Radcliffe-Brown 1930, p. 42).
The Serpent seems to represent simultaneously the powers of life
which is so vulnerable and death which is such a constant threat
in most of Australia.

As the element of water the 'Rainbow Serpent' cannot be identified with any particular water-site, but only as inhabiting the water-site of a region. Such is the case with the Waugal, the 'Rainbow Serpent' in or of the wetlands and rivers of south-western Australia where:

Waugal beliefs are widespread . . . and refer to a water-creative force with a serpentine physical manifestation. . . The Waugal is not *just* a mythic serpent, an Australian version of the Loch Ness Monster. The Waugal is not *just* a totemic ancestor. The Waugal is not *just* a spiritual being, a semi-deity. The Waugal is indeed all of these but is, more fundamentally, a personification, or perhaps more correctly animalisation, of the vital force of running water. As such, the question 'does this permanent river (or creek, or spring, or other water source) have (or belong to, or be associated with) a Waugal (or the Waugal) becomes, from an Aboriginal view point, meaningless and condescending. The presence of 'living water' bespeaks Waugal immanence. (O'Connor, 1989, p. 47)

Mircea Eliade has traced how 'living water, the fountains of youth, the water of life . . . is guarded by monsters' (Eliade 1958, p. 193).

Mudrooroo, though, maintains that the Waugal of south-western Australia is black and furthermore suggests that the idea of a 'Rainbow Serpent' common to all Australian indigenous cultures is an anthropologist's invention:

Watjelas [whitefellas] have studied us and have found that Aborigines all over Australia respect snakes, and they have joined up all these stories about snakes and made something called a Rainbow Serpent. They say and even tell us that Waugal is a rainbow serpent, whatever that is. But he isn't. He is a big, hairy snake that made the rivers and hills and valleys and then, after he had done this, went to sleep in the deep part of the river. If he is any colour, he is black, but when we tell them this, they say he is a Rainbow Serpent and refuse to listen. (Mudrooroo 1988, p. 33)[9]

The myth of the 'Rainbow Serpent', the myth of the myth, could thus be seen as having developed out of a drive to unify and homogenise the heterogeneous, even to imply that there must be some sort of originary and unitary ur-myth from which contemporary stories are derived, and of which they would be mere vestiges. Such a drive seems to be the product of a quest for origins rather than the result of a desire to understand the geographical richness and contemporary relevance of Australian indigenous cultures.

The living water of the wetland in Aboriginal cultures is worlds away and cultures apart from the dead water of the marsh, swamp, slough, etc., in patriarchal culture. But out of dead water, living water comes and without the latter the former would not be possible. Out of old life, the death and decay of the swamp, new life springs, the blossoms and fruits of the swamp. The 'Rainbow Serpent', Radcliffe-Brown continues, 'may be said to be the most important representation of the creative and destructive power of nature, principally in connection with rain and water' (Radcliffe-Brown 1930, p. 347). The 'Rainbow Serpent' is not a monstrous swamp serpent which only kills and consumes, but the Great Goddess of new life as well. The 'Rainbow Serpent' is, in the words of the Noonuccals, 'the giver and taker of life' (Noonuccal and Noonuccal 1988). Swamp waters are both life giving and death dealing – living black waters.

The combination of the giving and taking of life represents the mixing of birth and death in the swamp's ecology. Radcliffe-Brown argued earlier that the function of the 'Rainbow Serpent' is 'to maintain at their full power the forces of nature':

the processes of nature, the changes of the seasons, the growth, flowering, and fruiting of plants, the multiplication of animal species and of the human species itself, are not considered as just happening, but must be produced or provided by the society itself by means of the co-operative efforts of the various totemic groups amongst whom the various realms of nature are distributed. Now the 'Rainbow Serpent', whether regarded as a rainbow or a serpent, or species of serpent that lives in water-holes, is just as much a part of nature as the kangaroo or the sun or cold weather, and is therefore just as necessarily an object of ceremonies as these in any well-organised system. (Radcliffe-Brown 1926, pp. 23–4)

The 'Rainbow Serpent', as Kenneth Maddock put it more recently, 'in the water, but also in the sky' is not only 'responsible for female fecundity, but [also] responsible for destructive forces'. Indeed, the 'Rainbow Serpent' combines productive and destructive forces into one force, 'unites opposites in a totality', just as in the swamp's ecology. The 'Rainbow Serpent' is both male and female and both benignant and malicious, even malignant (Maddock 1978, pp. 2 and 8). But not exclusively malignant like the Stymphalian birds and sundry other marsh monsters and swamp serpents of patriarchal and filiarchal western culture.

Notes

1. For a further discussion of swarming, see Giblett 1993, pp. 541–59.
2. Similarly, the pre-Socratic 'Thales said that everything originated from water'. See Cicero 1972, p. 80. Yet according to Aristotle for Thales everything 'not only originates from but returns to water which is thus both eternal (and so divine) and the stuff of which everything consists'. *Oxford Classical Dictionary*, second edition, ed. N. G. L. Hammond and H. H. Scullard, Oxford, Clarendon Press, 1970, p. 1050. See also Aristotle, *Metaphysics* 983^b20.
3. For supine and panting as feminine attributes in patriarchy, see Threadgold, 1988, p. 50.
4. I am grateful to Sandra Giblett for pointing out the pertinence of this book.
5. I am grateful to Sandra Giblett for passing on this book to me.
6. See also Deleuze 1984, especially pp. 50–2.
7. I am grateful to Jon Stratton for pointing out the pertinence of this book.
8. Note the order of priority in the sub-title of Flood's book which renders indigenous people a subset of, object of ownership by and appendage to the land or country. Why not *The Story of Prehistoric Australian People (and Their Lands)*?
9. I am grateful to Hugh Webb for drawing my attention to this story. See also Mudrooroo 1994, p. 2.

V

Politics and Wetlands

Rebels and Runaways: The Wetland as Refuge and Site of Resistance

Wetlands have been seen as places of disease, depression, horror and the monstrous. Often as a direct result of this perception they have also been seen as a refuge for the runaway and a site of resistance for the rebel or revolutionary. Wetlands are easy to defend and hard to attack. For the theorist of war, von Clausewitz, 'low lands and morasses, if means of crossing are not too numerous, belong to the strongest lines of defence which can be formed'. Equally 'morasses, that is impassable swamps . . . present peculiar difficulties to the tactical attack' (von Clausewitz 1968 Vol. II p. 289; Vol III p. 29). The Dutch were to exploit these features of the Netherlands against the British, French and Spanish in the sixteenth and seventeenth centuries (Schama 1988, pp. 25–7). Wetlands warfare fulfils three out of five of von Clausewitz's criteria for waging a successful guerilla, or what he called people's, war: first, by being carried out in the interior of the country; secondly, by taking place in a theatre of war which 'embraces a considerable extent of country'; and thirdly, by occurring in country that is 'irregular, difficult, inaccessible', or of 'a broken and difficult nature', by reason of, amongst other things, marshes.

Though they are the interior of the country, wetlands may nevertheless border the exterior as in the case of the Fens and the Mekong Delta, in the sense of being seen as 'the impenetrable depths' where borders fold in on themselves to produce the ultimate invaginated interiority to be penetrated by phallic heroes. Furthermore, the successful guerilla wars waged in the Fens and Mekong Delta fulfil the additional criteria of being a theatre of war extending over a considerable area. Wetlands are also conducive to fulfilling Clausewitz's two remaining criteria: the war cannot hinge on a single battle, and the national character must support it (von Clausewitz 1968, Vol. III p. 343; Paret and Shy 1962, p. 12). Wetlands provide the opportunity for quick skirmishes, for stinging attacks and hasty retreats. They are also the place where national character has been forged and maintained. Again the Fens and the Mekong Delta are the obvious examples in both instances.

In this chapter I trace the history of wetlands warfare from Roman Britain to the Vietnam War, from pre-modern to post-modern times, and argue that many of the characteristics of the wetland which make it unattractive for the norm society, for example, its darkness and impenetrability, are precisely those features which make it ideal for the rebel or revolutionary during time of war or suitable for the runaway from oppression during the time of 'peace' in which there is a sustained deprivation of liberty. Depending on whose point of view, even whose side, is taken, the wetland is the locus of contradictory, mutually exclusive definitions, as is the definition of the rebel or revolutionary as either 'terrorist' or 'freedom-fighter'. Nowhere is this ambiguity more evident than when considering the role wetlands have played in military and slavery history where the same wetland can be on the one hand an intractable and inhospitable foe to the dominant but on the other a helpful and obliging friend to the dominated seeking refuge, and even a base for insurgency. In the Anglo-Saxon tradition specifically that double definition and history is figured most prominently in the Fens.

THE FENS AND OTHER ENGLISH MARSHES

English wetlands have, as Jeremy Purseglove pointed out recently, 'a long history as centres of resistance' (Purseglove 1989, p. 35). This long history stretches from the Romans and ancient Britons, through the Danes and Alfred the Great to the Normans and Hereward the Wake. In many ways this has been a predictable history because, as Purseglove goes on to observe, 'marshes have always been easy to defend'. That defence and long history first makes its way into the written record with the Roman invasion in which, H. C. Darby noted that the Romans 'encountered much difficulty' in their conquest of marshy regions (Darby 1974, p. 20). Xiphilinus related how in peace the Britons 'inhabit mountains wild and waterless, and plains desert and marshy' and how in war they are capable of 'hiding in marshes . . . many days with heads only out of water' (ibid. p. 20). The Romans, on the other hand, according to Dion Cassius 'wandered into the pathless marshes, and lost many of their soldiers' (ibid. p. 20). An embattled indigenous population on its home turf of a wetland was an intractable opponent as the Americans and their allies were to rediscover in the Plain of Reeds and the Mekong Delta. This was also particularly so with the Fens whose 'inhabited fields were isolated, their waters tidal, with great banks of treacherous mud, intricate and unbridged; such conditions are amply sufficient for a defensive war' as Hilaire Belloc surmises (Belloc (1906) 1970, p. 53).

Such conditions in the Fens were more than amply sufficient in the defensive war against the Romans who, according to Mary Chamberlain,

considered the area ungovernable and were outwitted by the cunning fen dwellers who could travel on stilts across the marshes, a skill which the Romans failed to master. The inaccessibility of the area made it a refuge for those fleeing from authority. The poverty and atmosphere of independence made it ripe for revolution and a stronghold of anti-authoritarianism. (Chamberlain 1975, p. 15)

The atmosphere of the Fens, which made it in one sense unhealthy from the prevailing medical point of view, also made it in another sense conducive to rebellion from a resisting political point of view.

One prominent Fen rebel was Hereward the Wake called the 'Last of the English' by Charles Kingsley. In his 'Prose Idyll' of the Fens, Kingsley describes 'how Nature left to herself, ran into wild riot and chaos more and more; till the whole fen became one 'Dismal Swamp,' in which the 'Last of the English' (like Dred in Mrs Stowe's tale) took refuge from their tyrants, and lived, like him, a free and joyous life awhile' (Kingsley 1884, p. 111). Kingsley devoted a whole book to *Hereward the Wake* which exemplifies again his ambivalence about the Fens. For Kingsley, 'they have a beauty of their own, these great fens, even now, when

they are dyked and drained, tilled and fenced – a beauty as of the
sea, of boundless expanse and freedom' (Kingsley 1908, p. 9). It is
this feature of the Fens in particular, the way in which for
Kingsley 'the green flat stretched away, illimitable' (ibid. p. 9),
which gives the Fens a large horizon of possibilities, especially for
defence. In the Fens in particular and the marsh in general, using
Belloc's words, 'one of the first requirements of defence is
afforded – an unbroken view of every avenue by which attack can
come. There is no surprising such forts' (Belloc 1970, pp. 53–4).

Such a fort was the Isle of Ely which Kingsley describes as 'a
natural fortress' (Kingsley 1908, p. 9). The precise feature which
makes the Fens such a good defence for the Fen defender and
such an object of aesthetic appreciation for the Fen dweller is
precisely, and hardly surprisingly, the feature which the cartographer and Sanitary Inspector bemoans, even vilifies. Yet for
Kingsley despite its obvious aesthetic and military virtues, 'that
fair land, like all things on earth, had its darker aspect. The foul
exhalations of autumn called up fever and ague, crippling and
enervating' (p. 10). Against the beauty of the Fens Kingsley poses
the miasmatic and melancholic diseases arising from its slime.

The lowlands generally, presumably including the wetlands, are
for Kingsley 'the soonest to be civilised and therefore the soonest
taken out of the sphere of romance and wild adventure into that of
order and law, hard work and common sense', and so into the
imperialist order of the straight lines of ditch, drain, dyke and
road. Yet despite being the soonest to be civilised they are also the
place of the backward and primitive: 'the lowlands and those who
live in them are wanting in the poetic and romantic elements',
another case of 'blaming the victim', the victim of 'civilisation', of
drainage and enclosure (ibid. pp. 1 and 2).

The lowlands also contrast unfavourably with the highlands, as
the civilised lowlands for Kingsley lack the mystery and sublimity
of the romanticised highland: 'there is in the lowland none of that
background of the unknown, fantastic, magical, terrible, perpetually feeding curiosity and wonder, which still remains in the
Scottish highlands' (ibid. p. 2). Yet Kingsley describes only a few
pages later how the uncivilised lowlands of the Fens were the
home of 'witches, ghosts, Pucks, Wills o' the Wisp' (ibid. p. 10).

This indigenous, pre-modern tradition of the fantastic and magical which fed curiosity and wonder in a satisfying diet for millennia in an autochthonous culture is counterposed unfavourably to the modern, Romantic sense of the sublime. Lowlands and highlands are counterposed, as are implicitly wetlands and drylands, and the high- and-dry-lands are valorised over and to the detriment of the wet- and-low-lands, the netherlands (see Chapter 2, Figure 1).

The dry highlands are the locus of the sublime whereas the wet lowlands are the site of slime, a feature which Kingsley dwells upon repeatedly in *Hereward the Wake*, especially when the slime is ostensibly 'bottomless' (ibid. pp. 282, 286) and certainly 'stagnant' (ibid. p. 421). The bottomlessness of slime contrasts unfavourably with the heights of the sublime. Yet it is this very same slime that is no less than 'the camp of refuge for English freedom' (ibid. p. 11) and the ultimate defence against the invading Normans who end up, according to Kingsley, 'drowning in the dark water', the black waters of malaria and melancholy, 'or more hideous still, in the bottomless slime of peat and mud' (ibid. p. 286), a fate worse than death as we saw also for heroic American soldiers in Vietnam.

In modern psychogeocorpography the seemingly bottomless and unplumbable depths of the slime of the wet lowlands (the interiority of the Body of the Mother) contrast unfavourably with the immense but measurable, terrifying but scalable, heights of the sublime of the dry highlands (the exteriority of the Law of the Father) (see Chapter 2, Figure 1). Similarly the clear, running, white water of the mountain stream conducive to bracing clarity of thought and flowing in demarcated watercourses contrasts favourably with the muddy, stagnant, black water of the wetland lying beneath the black sun of depression and melancholia and sitting still in labyrinthine b(l)ackwaters. Parts of the Fens are even called 'Black fen', 'black for the dark peaty soil; black – for the mood of the area, for its history and for its future', as Mary Chamberlain explains, even black for its waters (Chamberlain 1975, p. 11).

Rather than a staging ground for rebellion coming out of the wetland against a centralised authority, the wetland has been largely a site of resistance against the incursions and depredations of that authority into the wetland. Darby notes that

there were many local rebellions of both political and of economic origin in medieval England; but it is significant that the people of the fens never took part in any of these. . . . Their country certainly became a centre of resistance to authority, but to interpret that fact as the expression of any spirit of freedom fundamental in marshland peoples is quite unwarranted . . . The role of the Fenland in English history has been passive rather than active. It was repeatedly a region of refuge, rather than a breeding ground of freedom and discontent . . . the problem of fenland rebellion is not a social one with an economic background, but a military one with a topographical background. (Darby 1974, p. 143)

The passivity of the Fens' role in English history (his-tory) can no doubt be put down to its feminisation. History in general, though, has seen topography as background, as the stage on which events occur rather than as an actor in its own right. In order to give land (including wetland) a say, and part, in history, temporal history needs to be complemented by what Paul Carter calls spatial history (Carter 1987, intro.).

THE GREAT DISMAL, THE BLACK AND OTHER AMERICAN SWAMPS

A similar pre-occupation with the wetland as refuge and site of resistance was enacted on the other side of the Atlantic. Indeed, it is possible to argue that the tradition of the swamp as base for the freedom fighter and refuge for the renegade was transferred to the other side of the Atlantic, and then in the ultimate historical irony to the Mekong Delta during the Vietnam War. Even in the eighteenth century William Byrd, that chronicler of the Great Dismal Swamp, 'came upon a family of mulattoes that called themselves free, though by the shyness of the master of the house, who took care to keep least in sight, their freedom seemed a little doubtful. It is certain many slaves seek shelter themselves in this obscure part of the world' (Byrd 1966, p. 186). The Fens as the last refuge of English freedom is transformed into the American swamp as the first refuge of runaway slaves seeking freedom.

A century later Samuel Warner could still note of the Great Dismal Swamp that the 'darkest and deepest recesses' of 'this extensive Swamp has been for a long time the receptacle of runaway Slaves in the South' (Warner 1971, pp. 296–7). He could even go on to expatiate at length on the Swamp itself in typical miasmatic and melancholic terms which makes it nevertheless a perfect refuge for runaways:

neither beast, bird, insect or reptile, approach the heart of this horrible desert; perhaps deterred by the everlasting shade, occasioned by the thick shrubs and bushes, which the sun can never penetrate to warm the earth; nor indeed do any birds care to fly over it . . . for fear of the noisome exhalations that rise from this vast body of filth and nastiness. These noxious vapours infect the air round about, giving agues and other distempers to the neighbouring inhabitants . . . With all these disadvantages, the Dismal is, in many places, pleasing to the eye, though disagreeable to the other senses . . . It is within the deep recesses of this gloomy Swamp, 'dismal' indeed, beyond the power of human conception, that the runaway Slaves of the South have been known to secret themselves for weeks, months and years, subsisting on frogs, tarripins, and even snakes! (ibid. 297–8)

As the Swamp is inhospitable and uninhabitable to the white master because of its putative fevers and melancholias, it is the only place left for the runaway slave to hide. Precisely what makes the Swamp a wilderness in the pejorative sense for the white master is makes it the first, and only, refuge for the runaway.

In Harriet Beecher Stowe's *Dred; A Tale of the Great Dismal Swamp*, or more precisely, a tale of Dred, a runaway slave in the Great Dismal Swamp, 'the recesses of that wild desolation known as the 'Dismal Swamp',' are not only Dred's refuge but are also used as a figure for 'the dark recesses of a mind so powerful and active as his placed under a pressure of ignorance and social disability so tremendous' (Stowe 1856, pp. 216, 450). Stowe's narrator subscribes to the theory that as the swamp is isolated from the centres of official learning and the social circles of polite society, the swamp-dweller is inferior in intelligence and in the social graces compared to his or her dryland counterpart. The swamp is not only used as a figure for the mind of the runaway slave, but also there seems to be some sort of analogical correspondence between the socially deprived mind of the negro and the twisted goblin growths of the swamp.

Despite Stowe's anti-slavery stance, she uses stereotypically racist analogues. Such a correspondence is also to be found in the publisher's prefatory remarks to 'Nat Turner's Confessions', the basis for *Dred* (and Nat Turner the basis for Dred) and from which extracts are appended to it: 'a gloomy fanatic was revolving in the recesses of his own dark, bewildered, and overwrought mind, schemes of indiscriminate massacre to the whites' (ibid. p. 499). This runaway slave and slave revolt leader has a mind like a swamp according to the publisher of his confessions. In William Styron's reworking of *The Confessions of Nat Turner*, Nat is described as having been 'half-drowned from birth in a kind of murky mindlessness' (Styron (1967) 1981, p. 164). The mind of this black man infected by the black water of the swamp beneath the black sun of depression and melancholia contrasts strongly with what Stowe wryly calls 'the hot and positive light of our modern materialism' (Stowe 1856, p. 246) with its enlightenment and thermodynamic capitalist technology.

The swamp is both the stronghold of nature, the last wilderness, and the stronghold of rebellion easily defended by the runaway and revolting slave as Styron's fictionalised Nat Turner thought:

the Dismal Swamp ... [was] a perfect stronghold ...: though large ..., trackless, forbidding, as wild as the dawn of creation, it was still profusely supplied with game and fish and springs of sweet water – all in all hospitable enough a place for a group of adventurous, hardy runaways to live there indefinitely, swallowed up in its green luxuriant fastness beyond the pursuit of white men. (Styron (1967) 1981, p. 271)

The Swamp is hardly a Garden of Eden, but it is the work of the first day of creation which does not swallow the runaway in an act of oral sadism, but satisfies him or her orally in an act of generosity.

Rather than a Garden of Eden, Dred's 'fastness in the Dismal Swamp' (Stowe 1856, p. 403) is troped in Biblical terms as 'the stronghold of Engedi' (the title of Chapter 50) which is 'a freshwater spring on the west of the Dead Sea ... The fertility of the area in the midst of such barren country made it an ideal place for an outlaw, for food and hiding-places were readily available' (Douglas 1962, p. 369). Stowe's Biblical allusion is apt as 'amidst the wild and desolate swamp ... was an island of security, where

Nature took men to her sheltering bosom' (Stowe, p. 462). The bosom of Mother Nature for the runaway slave was, as it was for Nat Turner, 'the bosom of the Dismal Swamp' (Styron (1967) 1981, p. 80). For the runaway the Swamp is the good breast of the Great Mother.

Besides the runaway slave who used the swamp as refuge, the white revolutionary or rebel could use the swamp as base. Such was the case with the long line of 'swamp foxes', those revolutionaries or rebels, depending on your point of view, who used the swamp as base for amphibious guerilla warfare. Probably the most famous of these was Francis Marion who, according to Robert Bass, was 'barely able to read and write', yet was 'a tactical genius', 'a wily guerilla leader' and 'one of the most extraordinary heroes of the American Revolution' to boot, 'second only to George Washington' who held 'the vital woods and boglands of South Carolina' against the British from August 1780 to September 1781 (Bass 1959, p. 4). Marion belies Eric Hobsbawm's recent conclusion that before the First World War partisan style guerilla warfare was 'simply not part of the tool-kit of the prospective makers of revolutions' (Hobsbawm 1995, p. 79).

William Gilmore Simms, that poet of the edge of the swamp and also biographer of Marion under the title *The Partisan*, related how

the swamps of Carolina furnished a place of refuge to the patriot and fugitive, when the dwelling and the temple yielded none. The more dense the wall of briars upon the edge of the swamp, the more dismal the avenues within, the more acceptable to those who, preferring Liberty over all things, could there build their altars and tend her sacred fires, without being betrayed by their smoke. (Miller 1989, p. 86)

The swamp here is feminised as the good mother, liberty; indeed the swamp is metonymic with liberty. The swamp mother is the site of refuge when the fathers' institutions gave none. Like the open fen and marsh which afforded a view down every avenue by which attack can come and so made a fortified position easy to defend even by limited numbers, the closed swamp with its narrow avenues between its trees could provide a view of the approaching enemy and a quick and easy means of escape for outnumbered guerilla groups.

Besides serving as refuge and retreat, the swamp was also headquarters of operations for 'the swamp fox'. According to Noel Gerson, Marion 'developed a new type of warfare that the British had never before encountered', not only guerilla warfare of small-scale harassments and incursions, but amphibious guerilla warfare using the marsh and swamp as base and headquarters (Gerson 1967, p. 254). This type of warfare was what the United States was to encounter itself in Vietnam. Bass goes on to relate how:

bold and elusive, Colonel Marion was a haunting nemesis to the Tories, terrorising them from White Marsh to Black Mingo [Creek and Swamp]. He was an armed will-o'-the-wisp to the British soldiers, a phantom exacting retribution and justice. . . To the Carolina partisans Francis Marion was a latter-day Robin Hood. (Bass 1959, p. 3)

Marion is here represented as a kind of benign miasmatic emanation from the swamp, a complete turnaround from the usual associations of malaria, melancholia and monsters.

To these virtues of avenging angel, or perhaps in order to achieve them, Noel Gerson adds that Marion 'knew every inch of this vast, virtually uninhabited territory' of the marshes of the South Carolina seacoast (Gerson 1967, p. 5). The wetland could be used to tactical advantage in time of war whereas in time of peace it could and probably would revert to wasteland. For the rebel/revolutionary the wetland ceased to be wasteland because it was not surplus to requirements, but fulfilled an important military function. The wetland achieved in military terms a positive valuation (depending, of course, on whose side you were) which it did not tend to find in the wider culture, certainly not in peace, though perhaps herein lies the beginning of that fascination with the swamp in nineteenth-century American culture traced by David Miller in *Dark Eden*.

The American swamp has served three different military and political functions over the course of its recent history: firstly, as the first refuge of runaway slaves seeking freedom; secondly, as last resort and base for the white revolutionary struggling to overthrow imperial government; and thirdly, as the last refuge of indigenes trying to hold onto their freedom and lands. In the nineteenth century Charles W. Evers described how in the Black Swamp of Ohio 'the savage' was 'exempt from attack or pursuit' and therefore 'here enjoyed perfect freedom, and lived in accordance with his rude instincts and the habits and customs of his tribe . . . In war, this valley was his base line of attack, his source of supplies, and his secure refuge; in peace, his home' (McGarvey 1988a, p. 63). The swamp for 'the savage' was a homely, not an uncanny place.

In short, the prevailing view of nineteenth-century American culture was that, as Bruce McGarvey succinctly puts it, 'the Indian belonged in the swamp. The white man belonged outside the swamp' (ibid). Presumably in the south, the black man 'belonged' not in the swamp as runaway but outside the swamp as slave with the white master. The white man only belonged in the swamp in the extenuating circumstances of a revolutionary cause when hard pressed by the oppressor and so when the swamp was the last resort. The swamp gave birth to the freeman and freewoman from the slave and to the revolutionary from the citizen.

One group of Amerindians who made the swamp their home and successfully resisted the white invasion in it for a time were the Seminoles of Florida for whom, as George Buker puts it, the 'vast swamplands' of the region served both as 'a refuge and base of operations' (Buker 1975, p. 5). According to Buker, 'seminole' means not 'runaway slaves' though they did join, but 'runaways', who, as Peter Tremayne puts it, were renegades or separatists from the Creek confederation (ibid. p. 7; Tremayne 1985, p. 96). Two 'wars' against the Seminoles were fought, the first in 1817–18 and the second, a protracted affair from 1835–42 during which, according to V. G. Kiernan, '30,000 troops were put into the field and more than US$20 million were spent' (Kiernan 1978, p. 30).

The second 'war' ended indecisively when, Buker relates, 'many of the Seminoles remaining in the territory withdrew to the Everglades in South Florida to wage a last-ditch fight using guerilla tactics. The Seminoles were reduced to around 300 by 1842 when [the government in] Washington finally acknowledged that it would be impossible to track down every Indian hiding in the swamps' (Buker 1975, p. 14). Although the Seminoles were outnumbered by one hundred to one, they still 'won' the war because of its location. As John K. Mahon indicates, 'swamp land was one of the terrain features which gave peculiar and strenuous character to the Seminole War' (Mahon 1967, p. 129). One white settler remarked of the terrain in 1835 that it was 'the poorest country two people ever quarrelled over . . . It is a most hideous region to live in, a perfect paradise for Indians, alligators, serpents, frogs and every other kind of loathsome reptile' (ibid. p. 134). Indians are lumped together and equated with loathsome reptiles.

The features of this terrain helped to develop the characteristics of one strain of what Buker calls 'riverine warfare' in most cases of which the phallic 'thrust into the enemy's land' through the 'fluid concourses' of 'interior waterways' of the Body of the Mother, including 'vast swamplands', would be 'met by military rather than naval resistance, and the major confrontations would not be naval' (which is hardly surprising given that the Seminoles had no navy). Yet the war was fought as much, if not more, in the water as on the land. 'Therefore', Buker concludes, 'forces must be specially trained combat groups organised for sustained operations in both elements', presumably the elements of both earth and water though Buker does not specify (Buker 1975, p. 5). The white forces were poorly prepared for such operations as the mixed elements of the wetland were not suitable for 'the European tactics of massed formations and decisive engagements'.

Despite this drawback the early commanders pursued this strategy by sending, as Buker puts it:

large columns of regulars and militia into the wilderness with complex plans to converge upon Indian strongholds and defeat them on the field of battle. The unfamiliar Florida terrain hampered, delayed, and discouraged the American military columns as they thrashed about building roads, fording rivers, erecting forts, and trying to create a supply system to support their movements. (ibid. pp. 13–14)

These strategies used the tools and techniques of modernity which were so successful for adventure heroes but, of course, were unsuccessful for ordinary soldiers against the tactics of the highly mobile Seminoles who had an intimate knowledge of the wetland.

For the white soldiers the swamp was a living hell on earth as Mahon relates:

it is indeed doubtful if United States ground forces endured harsher field conditions anywhere. Protracted service in the humidity, the rank growth, and the darkness of the Florida swamps took the sunshine out of a man's life. Cypress knees, mangrove roots, and sawgrass tortured the foot soldier. Too much water, and the lack of water, made his life a torment. There was marching in water from ankle- to armpit-deep, hour after hour, with no chance to dry off, not even at night . . . Cut by sawgrass, made raw by insect bites, now and again reeling dizzy from dysentery and fever, the common soldier in Florida lived in a world which had no horizons. (Mahon 1967, p. 240)

For the Seminoles the world of the wetland was the horizon of their world which they knew and understood. Similar conditions were to be encountered by the American soldier over one hundred years later in Vietnam; similar lessons about how to fight amphibious guerilla warfare were not to be learnt. Mahon concludes that 'the Second Seminole War is important in American military history because of its development of guerilla, or partisan-style, warfare' (ibid. p. 325). This style of warfare was later employed by the Vietnamese against the Americans.

THE MEKONG DELTA, THE PLAIN OF REEDS AND THE VIETNAM WAR

Many of the military and political functions served by the Great Dismal Swamp and the Fens and many of the meanings produced in relation to them were repeated in an uncannily similar manner with the Mekong Delta and the neighbouring Plain of Reeds (*Plaine des Joncs*) in Vietnam. The Mekong Delta has been described as 'a vast, underdeveloped and underpopulated frontier land' (Michigan State University Vietnam Advisory Group, p. 1) and the Plain of Reeds has been called 'a great swampy area' (Tanham 1967, p. 91). Together these two areas make up the area formerly known as Conchinchina (Sansom 1970, p. 5). After the Mekong has flowed through what British geographer Paul Lightfoot (a good quality in a student of wetlands) calls 'flat, desolate, almost featureless country', the river terminates in the Delta made up, like the Fens, of 'flat, featureless and almost desolate lands', or perhaps more precisely wetlands (Lightfoot 1981, p. 56).[1]

Desolate and almost featureless the land on either side of the Mekong River approaching the Delta may be from an aesthetic and

geographical point of view, and featureless and almost desolate
the wetland of the Delta itself may be from a similar point of view,
but from an agricultural and economic perspective it is quite
different. According to Joseph Buttinger 'the delta of the giant and
benign Mekong River' is 'vast and fertile' (Buttinger 1958, pp. 35,
48). In fact, the name Mekong evidently means 'the grand river
mother' (Honda 1972, p. 113). Similarly Buttinger anthropomorphises the river by suggesting that 'compared to the Red River,
the Mekong could almost be called benign' (Buttinger 1972, p. 9).
If the River is benign, the Delta is downright beneficent. The
Mekong Delta is a kind of big momma wetland/womb for
Buttinger who refers to it repeatedly as 'the vast and fertile
Mekong River Delta' (Buttinger 1967a, p. 520, n. 3).[2]

Buttinger's insistence on invariably referring to the Delta as the
Mekong River Delta privileges the River by seeing it as the creator
of the Delta. Perhaps it is in terms of a patriarchal western
teleological logic (the river runs to the sea, the river builds the
delta into the sea (Buttinger 1972, p. 9)), but it is not in terms of an
eco-logic which sees the delta as a dynamic wetland ecosystem of
interchanges between river and sea at once building from the
former to protect against the incursions of the latter which is
filtered to provide water for the land and so nourishes it like the
placenta. Despite the fact that it proved – disappointingly from
the point of view of French colonialism, thankfully so from the
Vietminh and Viet Cong side – 'largely unnavigable' and 'no trade
route,' Buttinger prefers the Mekong River as the good breast
whose waters 'irrigated Vietnam's richest rice lands with pleasing
regularity, in addition to millions of acres waiting to be opened up
for the production of crops big enough to make Vietnam one of the
world's biggest exporters of rice' (Buttinger 1967a, pp. 9, 19).

The River is the good breast which waters regularly the Delta
which, in turn, is the virgin (wet)land passively awaiting the
'opening up' and penetration of patriarchal agricultural technology to inseminate its fertile womb and bring forth gigantic (but not
monstrous) progeny. They are created in the image of the father,
but also in the likeness of the mother delta since the River
culminates in a delta of 'approximately 45,000 square kilometres
of mostly alluvial soils' (Duiker 1983, p. 1).[3] Given the richness of
the soil and the abundance of water, it is hardly surprising that the

Delta is 'Vietnam's granary', or perhaps more aptly 'the country's rice basket' (Fall 1967, p. 384; Burchett 1965, p. 179). It is Mother Earth's 'bread basket'.

If the Mekong is the great river mother, then the land of Vietnam has been troped as her young, sway-backed daughter with her 'narrow waist of Central Vietnam' (Duiker 1983, p. 1). Extending the bodily metaphor in sagittal section, the Mekong River is the alimentary canal of much of Indochina which draws nourishment from its interior to fertilise 'the nether regions' of the Mekong Delta tucked away underneath its rump of the south. It is from here that new life springs so the Delta is both uterus and vagina rather than bowel and anus. Perhaps the Delta is more precisely the cloaca of Vietnam which serves both reproductive and evacuative functions. The Delta is both fertile and fertilising.

Like the Fens, the Mekong Delta and the Plain of Reeds have been subject to various attempts to canalise and drain them and to establish agriculture on their fertile plains. Yet the French drainage program may have in fact been the result of previous, alternative and unsuccessful attempts at subduing rebellion in Conchinchina which date back to the early 1860s (Long 1973, pp. 11–12). Drainage, as we saw in a previous chapter, is a colonial device for subduing an ostensibly recalcitrant, even rebellious, indigenous population and wetland environment. Typically the canalisation and drainage of the Mekong Delta followed the European model as Malcolm Browne points out: 'Delta canals in Viet Nam, all of them man-[sic] made, are straight as arrows. Even from the air, they stretch all the way to the horizon, like glittering slashes across the verdant plain' (Browne 1965, p. 12).[4] Straight canals, roads and railways through wetlands are imperialist devices, as are their representation in the straight lines of maps.

If the French drainage programme was designed to subdue rebellion, then it proved not only unsuccessful, but also counterproductive. This was certainly the case in November 1940 with the Plain of Reeds insurrection which was brutally suppressed by the French (Buttinger 1967a, pp. 225, 567, n. 129). The Plain of Reeds and the Mekong Delta were to continue to be a site, and seat, of resistance to the French over the next fourteen years. When the French returned to Vietnam after a brief Japanese interregnum in 1945 one of the most difficult areas to subdue was, according to Buttinger, 'along and beyond the Mekong River' and 'it took them many weeks to cover the difficult terrain between the major cities of the south' (ibid. p. 336).

A little later the Plain of Reeds was 'to become famous . . . during the [First] Indochina War [of 1946–54] as a Communist base' (ibid. p. 225). In his book on the conflict entitled aptly (for a wetland war in part) *The Quicksand War*, in the first part (again aptly) called 'Bogging Down' and in a section (also aptly) headed 'The War in the Mud' Lucien Bodard includes the account of a French officer who had just returned from patrol and who is reminiscent of the gung ho officer in Kafka's short story 'In the Penal Colony'. His account reproduces many of the stereotypical images associated with fighting an amphibious guerilla war. Bodard relates how: 'two hours earlier he had been worn out, covered with mud from head to foot. Now he was a pink cherub, kind and smiling. He was a well-bred young man, and in order to give me pleasure he told me about his expedition in detail':

My objective was a Vietminh 'factory' on a little island in the Plain of Reeds. Usually I put my men into sampans and we go up the canals, whose banks are lined with water palms so that they hem you in like the walls of a prison. You can't see anything, but you can be seen. This time, so as not to be picked up, we went on foot through the shit, as we call it – the stinking marsh covered with reeds and lotus that stretches out forever.

The approach lasted for hours, and all the time we were up to our waists in the muck. We had started at midnight: by dawn we had still not been seen. We were only a few hundred yards from the 'factory'. Then came the horn sounding the alarm. I looked through my field glasses. The little island was like an ant hill that has just been stirred up. I heard explosions and saw flames. I knew what that meant. The Viets were taking their heavy equipment away in boats and they were carrying out their scorched-earth policy with the rest. It always happens that way.

It took us another half hour to get ashore, to struggle up on to the firm ground. There was nothing but destruction and emptiness . . . We did not see a single man . . . There were certainly hundreds of Viets still there, but so hidden that there was not a hope of finding them. Some would have turned into bushes. Others would be right down under the mud, breathing through hollow bamboos. Most would be in carefully prepared burrows. It was like being surrounded by a crowd of ghosts: they would stay there for hours on end, waiting for us to go. To have unearthed them, we should have had to spend days there, searching everything. We couldn't do it. It would have been too dangerous. Besides the Legion needed us elsewhere for other assignments. . .

I must have been on hundreds of operations here in the Plain of Reeds. It is unbelievably monotonous. There are the ambushes – the ones you set and above all the ones you fall into. Suddenly men you can't see, amphibious creatures in the water and the mud, start firing: you don't even know where the shots are coming from. Then there is the chasing. Sometimes it is we who are after the Viets and sometimes the other way about, but its is always the same weird business. Men sunk deep in the mud, the gluey mud, slowly forcing their legs through it to catch other men who are bogged down in just the same way. You only see their heads, and the Plain of Reeds is so flat that you might think that they were dots stuck to its surface. When you come to the open space of a canal or an irrigation ditch you swim under the water, so as not to present a target: all that you see of a man is one hand holding up a tommy gun or the part of a mortar that must never get wet. Often some of my men are drowned.

Our great weakness is the wounded. I don't know what the Viets do with theirs: they never seem worried by them. But as soon as one of my men is hit I have to call off whatever operation we are carrying out. To get him out through those everlasting marshes I need every man I have – eight to carry him and the rest as a guard. It all turns into a dangerous, slogging retreat. If there are a lot of Viets behind us it can turn very nasty, seeing that we have this dead weight – all the more so since at the best of times they get along faster than we do. (Bodard 1967, pp. 38–40)

After all, this is their 'home ground'. If the French had an informer they could surprise the Vietnamese when the tide was on the ebb, but then the mud was no cover, despite their attempts at 'sinking into the water and the mud of the submerged forest and wriggling in among the mangrove roots'. A massacre ensued as the French officer went on to relate proudly, and terrifyingly (ibid. p. 40).

As Francis Marion the partisan, even the American Partisan par excellence, held the boglands of South Carolina in 1751 during the American War of Independence against the British, so nearly two centuries later in 1947 'Vietminh partisans prowled the Plain of Reeds' (ibid. p. 7) of South Vietnam in the war of independence against the French. Yet these guerillas perhaps not so much prowled as hid. According to Duiker, 'during 1947, the Vietminh were 'reduced to a struggle for sheer survival . . . [as] guerilla units hid in . . . the Plain of Reeds' (Duiker 1983, p. 43). Here they literally had to 'lie low' in the concretisation of a stereotypical metaphor used to describe guerilla tactics (Hunt 1974, p. 58).

Despite having to lie low and to struggle for sheer survival, the Plain of Reeds remained guerilla-controlled (Buttinger 1967b, p. 741). This control remained unchallenged there, as it did in the Mekong Delta despite some limited French success here in late 1949, during the entire course of the war (ibid. p. 973; Buttinger 1973, p. 91; Lockhart 1989, p. 212). No doubt such unchallenged control was due in no small measure to the wetland theatre of war in which the French had to fight with much difficulty and to their considerable chagrin, disgust and ultimate defeat. Despite the changes of drainage and despite the fact that in its wake the Mekong Delta became, according to Lightfoot, 'one of the world's most important rice-exporting areas by the 1930s, parts of the delta remained empty and inaccessible, providing ideal places for guerilla soldiers to hide out during the Vietnam War' (Lightfoot 1981, p. 59). Lightfoot's description of parts of the delta as 'empty' begs the question of empty of what?

Yet Wilfred Burchett argues that the Delta was far from ideal for guerilla insurgency. In comparing a district in Central Vietnam with the Delta, Burchett argues that although the former 'had the most unfavourable conditions for guerilla warfare in Central Vietnam, it was a paradise as compared to the flat Mekong Delta, with no mountains at all and forest only in the unpopulated mangrove swamps bordering the coastal areas' (Burchett 1965, p. 192). A similar, general view is to be found in Eric Hobsbawm's recent proclamation that 'guerilla warfare is most easily maintained in bush, mountains, forests and on similar terrains' (Hobsbawm 1995, p. 81).

Since when can guerilla insurgency take place most successfully only in the mountains and the forest? On land, in other words? What about in wetlands, as the Vietminh demonstrated? What about 'aquaterrains'? The Viet Cong were not just guerillas, but what a caption to a photograph in Lightfoot's book calls 'underwater rebel guerillas' (Lightfoot 1981, p. 59). Or more precisely from their point of view amphibious revolutionary guerillas.

Despite, or perhaps because of, the drawbacks of the Delta for a western-style land-based campaign, or even land guerilla warfare, Burchett argues that 'the Delta peasants . . . are the most experienced and probably the best guerilla fighters the world has ever known', perhaps not only because they have, as Burchett puts it, 'been raging almost continuous warfare for nearly a quarter of a century', but also because they could fight equally well in wetlands as on the land (Burchett 1965, p. 229). The Delta was the scene for wetland guerilla, or more precisely revolutionary, warfare, just one recent instance in a long line of defensive wars, peasants revolts and struggles staged, and waged, in and over wetlands.[5] By the early 1960s resistance in the Delta had been going on, on and off, for at least one hundred years.

For both sides in the Vietnam, or more precisely Second Indochina, War the Delta was strategically vital. According to George Tanham, some observers even suggested that 'Viet Cong strategy has as its primary objective control of the Mekong Delta' whilst the United States and South Vietnam forces stated that 'the Delta is the key to the war' (Tanham 1967, p. 150). Yet from the very earliest moments of United States involvement in Vietnam there was a recognition that the Mekong Delta would prove problematic as a theatre of operations. When General Maxwell D. Taylor recommended the introduction of US military forces to South Vietnam in a cable to President Kennedy in 1961 he stated that 'as an area for the operations of US troops, South Vietnam is not an excessively difficult or unpleasant place to operate . . . The most unpleasant feature in the coastal areas would be the heat and, in the Delta, the mud left behind by the flood' of that year (*Pentagon Papers* 1971, pp. 147–8).

As in the First Indochina War, 'the rice and population heartland of the Mekong Delta' remained in Viet Cong control despite concerted attempts at dislodging them, some of which were nearly successful (ibid. p. 250). Wilfred Burchett relates how in the Delta 'during 1962 helicopter-borne troops took a fairly heavy toll of resistance fighters and there was a period when [the National Liberation] Front [Viet Cong] leadership almost decided the price was too high, that resistance should cease and regular Front armed forces should withdraw to bases in the mountains' (Burchett 1965, pp. 192–3). Burchett was then informed immediately by one of the military leaders that 'when we discussed this . . . we realised in the bottom of our hearts that to withdraw from the Delta would mean never to return. It would mean to abandon the most revolutionary region, the foyer of the first uprising against the French in November 1940 [the so-called Mekong Delta Uprising of 1940] and of the first resistance war in 1945' (ibid.)[6]

Perhaps these and later attempts at dislodgment were nearly successful because the Commander in Chief of US forces in south Vietnam, General Westmoreland, seems to have been a student of US military history, especially of the Second Seminole War, as he recognised the distinctive geographical features and military requirements, especially transportational problems, of the Delta. In a memorandum to Pacific Command in 1967 he indicated that:

emphasis will be accorded the opening and securing of principal water and land lines of communication which are the key to all operations in the Delta. It is noteworthy on this score that effectiveness of forces available is hampered by an inadequate mobile riverine force . . . The Mekong Delta Mobile Riverine Force originally was tailored and justified as a four River Assault Squadron level. This requirement still is valid. The primary media of transport in the Delta are air and water. Air mobility is recognised as critical to success of operations in the area, but the size of offensive operations that can be mounted is limited by the inherent physical limitations of airborne vehicles. Accordingly, any sizeable offensive operations . . . must utilise the 300 kilometres of waterways in the Delta to exploit tactical mobility. (*Pentagon Papers* 1971, pp. 571–2)

Yet what use was the supposedly superior strategic mobility of helicopters, planes, and fibreglass dinghies, and even M113 armoured, amphibious personnel carriers, against the hit and run tactics of amphibious or 'underwater' guerillas who moved around readily in or on the mud?[7] Very little. The M113 was 'a five ton monster', a wetland war machine monster like the dredger for Davidson and Swift, which was 'supposed to be equally mobile on land, on water, or in mud' but which 'runs very well on dry ground or in deep water, but has trouble in mud of a certain depth', in other words, in wetlands, either the natural ones of the Delta or the artificial ones of paddy fields. The Viet Cong, by contrast, 'for years had used little one-man [*sic*] boats shaped like a saucer' which were propelled with one leg and which could 'scoot across the gooiest mud' (Browne 1965, pp. 9, 13).

The Vietnam War for Frederic Jameson was the 'first terrible postmodern war' because it 'cannot be told in any of the traditional paradigms of war novel or movie' given the fragmentary and chaotic nature of the experience (so brilliantly captured in Michael Herr's *Despatches*), though this was so much grist to the mill of television news with its live crosses and ten-second grabs (Jameson 1984, p. 84). No doubt its televisibility was part of its postmodernity. Yet it was also the first postmodern war because it was afflicted with a profound sense of *déjà vu* and irony. The heaviness and ironies of history hung over its head about to fall with the inevitability of defeat, with this time the United States on the losing side.

The war was postmodern in that from the American side it was ostensibly not a war of geographical conquest, of expansionism, nor a war of military defence. Rather it was one of ideological defence, even ideological and capitalist offence. It was also a postmodern war because it came at the tail end of a whole series of wars, resistances and revolts fought in large part in and over wetlands. So it was burdened with even a more acute sense of *déjà vu* and irony for the United States since here was the Viet Cong doing exactly what the second most famous hero of its own revolution had done – holing up, and winning out, in wetlands. History had in a sense come full circle to turn back on itself. If there is a postmodern 'end of history', it is because history is now repeating itself with a difference.

Although Westmoreland knew the lessons of history, particularly of the Second Seminole War, he did not know that other lesson of history, that they are never learned. Despite the best attempts and efforts at avoiding it, history always repeats itself with a difference. Whereas the Seminole defence was brought to a standstill or stalemate, the Viet Cong were victorious. The simplest explanation for the United States' different sorts of loss on both counts is that their forces fought the Seminoles 'at home', for both over mutually contested geographical territory, the national territory for both, whereas in Vietnam they fought the Viet Cong in their own (wetland) home and national territory a long way from mainland United States. Whereas for the Viet Cong it was a war of revolution against a colonial regime and a war of defence against a foreign invader and an imperialist power, for the United States it was always a war of ideological defence and imperialist and capitalist offence fought out on a foreign battlefield. Although Westmoreland knew the theory to prevent the repetition of history in principle, this did not empower him to prevent it in practice.

And last, but by no means least, the Vietnam War was a, if not the first, postmodern war because it was fought in part in the postmodern wetland of the Plain of Reeds and the Mekong Delta. This vestige of the pre-modern wilderness had in part been transformed into modern productive agricultural land through canalisation and drainage, to become the site of absentee land-owners and displaced peasants and so a hotbed of revolution, and in part been left to remain a modern wasteland outside of agriculture, had then as a whole agricultural and non-agricultural area been made over into a postmodern wilderness in military terms, a theatre of war for rebels, revolutionaries and counter-insurgents. For the United States, as for the French before them, the Delta was a postmodern wilderness of lost lives, wasted resources and failed large-scale industrial technologies and military strategies. Buttinger relates how 'one single amphibious operation in the Mekong delta was said to have cost [the United States] $16 million' (Buttinger 1972, p. 110). However, for the Viet Cong the Delta was a postmodern wilderness of refuge and site of resistance, revolutionary struggle, guerilla tactics and small-scale, hand-made technologies.

The wetland of the Mekong Delta was postmodern because it was the site of contesting, indeed contradictory, meanings. Similarly, the wetland in general is a place of double meanings, not simply of ambiguity, that keyword of modernism for Franco Moretti, (Moretti 1988, p. 340) but of ambivalence, the keyword of postmodernism. In postmodern times the wetland is literally ambi-valent, a place of moving values, its meanings circulating and swirling around, like its waters sometimes do, never settling on one definition, never flowing to an end point. At once the wetland is wasteland to be filled or drained, turned into profitable agricultural land or into a sanitary land-fill site and at the same time it is wilderness not yet subject to a capitalist imperative in which to seek refuge and from which to mount resistance to cultural and military imperialism.

NOTES

1. I am grateful to Ann-Marie Medcalf for drawing the section on the Mekong Delta in this book to my attention and for generally pointing out the pertinence of the Mekong Delta.
2. See also Buttinger 1972, p. 9 for a slight variation.
3. According to Buttinger the area of the Delta is 8,500 square miles, about half the size according to Duiker. See Buttinger 1972, p. 10.
4. The 'new face of war' for Browne is what he calls 'Paddy War' fought by the Viet Cong in rice paddies which elicited the response of what he calls 'mechanised paddy war' from the governments of South Vietnam and the United States. Much of the Vietnam War was arguably a kind of wetlands war fought out in what Browne calls 'marshy rice fields'.
5. Bernard Fall draws a useful distinction between guerilla warfare, literally 'a little war' of small-scale raids, and a revolutionary war which 'swept away the existing system of government'. For Fall, the Vietnam War falls into the latter category. See Fall 1963, p. 356.
6. Burchett concurs with his informant about the revolutionary status of the Delta and reiterates that it was 'the most revolutionary area in all of Vietnam' (p. 228). For the Mekong Delta Uprising of 1940, basically a peasant revolt, see McAlister 1969, pp. 131–3.
7. For an elaboration of the distinction between strategy and tactic, see de Certeau 1984, p. xix.

10

Conclusion: Writing a Word for Wetlands

Against the rhetoric of swamps and marshes as places of melancholy and monstrosity, of horror and disease there has been a counter-tradition in the patriarchal west which has regarded the wetland as a sacred place, a place of both death and life. Henry David Thoreau, John Muir and Aldo Leopold – the prominent precursor, modern founder and postmodern upholder of the American conservation movement – are the leading lights of this counter-tradition. These three wilderness wanderers not only 'spoke a word for Nature' in general as Thoreau put it, but also wrote a word for wetlands in particular (Bode 1982, p. 592). Yet the counter-tradition by no means begins with them, nor belongs to them exclusively – Sidney Lanier in the nineteenth century wrote lovingly of American swamps and marshes – nor ends with them as the recent, Nobel prize-winning poetry of Seamus Heaney about Irish bogs shows.

If it is possible to find a beginning for the counter-tradition in patriarchal western culture, it is perhaps with Adam Seybert who, though a subscriber to the miasmatic theory of malaria, wrote nearly two hundred years ago that 'heretofore mankind seems to have viewed the existence of marshes as noxious to them and unnecessary to their happiness ... I am of opinion that ere long marshes will be looked upon by mankind as gifts from Heaven to prolong the life and happiness of the greater portion of the animal kingdom' (Seybert 1799, p. 429). Indeed, wetlands are arguably the first gift of Heaven. Although 'mankind' may not yet have adopted this view, at least western or westernised 'mankind', it was only some fifty years later that Thoreau came to

228

see swamps as essential to his own life and happiness, though Seybert may not have included 'mankind' in the animal kingdom.

HENRY DAVID THOREAU: PATRON SAINT OF SWAMPS

If swamps could be said to have a patron saint, it would be the nineteenth-century philosopher and naturalist Henry David Thoreau who maintained that 'when I die you will find swamp oak written on my heart' (Thoreau 1993, p. 17). The swamp not only wrote on Thoreau; he also wrote repeatedly on it, a fact not often noticed or remarked upon by the literary critics or writers on the American idea of modern and postmodern wilderness (Nash 1982; Oelschlaeger 1991).[1] One of the few critics to do so is Carl Bode who refers to Thoreau's 'love for swamps' and how 'he enjoys being in them, enjoys writing about them' (Bode 1982, p. 686). Unlike Byrd and Sartre who found slime an impossible writing surface and feared slime writing on them, Thoreau loved to write on the swamp, and be written on by the swamp, or at least by the swamp oak.

Thoreau touches on and counters, or subverts, many aspects of standard swampspeak which we have encountered previously. Although Thoreau by no means addressed, and countered, every point of standard swampspeak since he seemed to be more concerned to produce affirmative press for swamps rather than simply rebuff, or even rebut, the pejorative, he did, nevertheless, turn the rhetoric against itself. He rejected the miasmatic theory of disease by stating that 'miasma and infection are from within, not without' (Thoreau 1980, p. 261). He countered the theory by suggesting that 'the steam which rises from swamps and pools is as dear and domestic as that of our own kettle' (Thoreau 1982, p. 61). For Thoreau, swamps and stagnant pools were not the antithesis of, nor a threat to, the domestic, but of comparable value. He did not valorise the wetland over the domestic but gave them equal value unlike those of his (and my) contemporaries who denigrated and feared the wetland (and accordingly valorised the canny over the uncanny).

Rather than seeing the airs of swamps as bearers of disease, Thoreau made a crucial distinction between fog and miasma, and even saw fog as healing:

The fog . . . in whose fenny labyrinth
The bittern booms and heron wades;
Fountain-head and source of rivers . . .
Spirit of lakes and seas and rivers, . . .
Bear only perfumes and the scent
Of healing herbs to just men's fields!
                              (ibid. p. 166)

In a poem devoted exclusively to the subject and entitled simply 'Fog' Thoreau referred to it as 'dull water spirit – and Protean god', as 'incense of earth', 'spirit of lakes and rivers' and as 'night thoughts of earth' (ibid. pp. 237, 238). Rather than a vector of disease and a cause of death like miasma, fog is a source of new life. Rather than regarding fog and mist as vapours bearing disease and death, why not see them as the visible manifestation of the exhalations of the earth, particularly of the trees, on which we are dependent for life? After all, we are in symbiosis with the oxygen-producing plants of the earth.

The swamp vapours were as domestic for Thoreau as kettle steam because the swamp itself was better than a domestic garden. Indeed, if Thoreau had had to choose between them he would have chosen the swamp every time: 'yes, though you may think me perverse, if it were proposed to me to dwell in the neighbourhood of the most beautiful garden that ever human art contrived, or else of a Dismal Swamp, I should certainly decide for the swamp' (ibid. pp. 612–13). Why? Because 'I derive more of my subsistence from the swamps which surround my native town than from the cultivated gardens in the village' (ibid. p. 612). The swamps are 'the wildest and richest gardens that we have. Such a depth of verdure into which you sink' (Thoreau 1962, IV, p. 281). Thoreau was no mere walker by the wetland, but a wanderer in the wetland who was not afraid of sinking into it as long as he eventually found a hard bottom.

Rather than the Garden of Eden, for Thoreau 'some rich withdrawn and untrodden swamp . . . is your real garden' (Thoreau

1993, p. 198). Yet this preference for swamps over town gardens was no mere nostalgia for a pastoral paradise lost as 'hope and the future for me are not in lawns and cultivated fields, not in towns and cities, but in the impervious and quaking swamps' (Thoreau 1982, p. 611). Thoreau's rhetorical tactic against the stratagems of standard swampspeak was to displace and upset the usual or normative disjunction between swamp and garden by seeing the swamp as garden, and so exploit the favourable associations of the garden as a place of light and life.

Thoreau also upset the usual dissociation between swamp and city by seeing the city as swamp and so subverted the unfavourable connotations of the swamp as a place of darkness, disease and even death. Rather than the swamp, Thoreau saw 'society', 'civilisation', and the modern city as bearers of disease, or perhaps more precisely he saw the modern city as swamp in the conventional sense of an uncanny and unhomely place of disease and horror, and saw simultaneously the swamp as canny and homely, as postmodern dwelling in the unconventional sense of a domestic, but also wild, (domestic *because* wild) place. Thoreau could 'see less difference between a city and some dismallest swamp then formerly. It is a swamp too dismal and dreary, however, for me'. Although he would prefer the swamp as swamp over the city as swamp, he nevertheless goes on to make a finer distinction: 'I would prefer even a more cultivated place, free from miasma and crocodiles' (Thoreau 1962, II, p. 47).

The city as swamp, however, had its own diseases and horrors:

let us settle ourselves, and work and wedge our feet downward through the mud and slush of opinion, and prejudice, and tradition, and delusion, and appearance, that alluvion which covers the globe, through Paris and London, through New York and Boston and Concord, through church and state, through poetry and philosophy and religion, till we come to a hard bottom and rocks in place which we call *reality*. (Thoreau 1982, p. 350)

Thoreau's homoerotic search for a hard or tight bottom has been remarked upon by a number of critics.[2] What has been less remarked upon is the fact that the hard bottom is primarily that of a pond or swamp, though 'there is a hard bottom everywhere', even 'with the bogs and quicksands of society' (ibid. pp. 568, 569).

Thoreau could 'fancy that it would be a luxury to stand up to one's chin in some retired swamp a whole summer, scenting the wild honeysuckle and bilberry blows, and lulled by the minstrelsy of gnats and mosquitoes!' (ibid. p. 187; Thoreau 1962, I. p. 141) Such music is the counterpoint to modernity: 'in all swamps the hum of mosquitoes drowns this modern hum of society' (Thoreau 1962, I, p. 83). Standing up to one's chin in a swamp was also a good place to smell the odours of the swamp, 'a strong and wholesome fragrance' of the vegetation 'by overgrown paths through the swamp' (ibid. IV, p. 478). He even said 'I love the smell of the swamp, its decaying vegetation' (ibid. IV. p. 305). Thoreau valued the sense of smell over the sense of sight to the point that 'methinks the scent is a more primitive inquisition than the eye, more oracular and trustworthy . . . The scent reveals, of course, what is concealed from the other senses. By it I detect earthiness' (ibid. IV p. 40). Indeed, for Thoreau 'that peculiar fragrance from the marsh at the Hubbard Causeway' was 'the fragrance, as it were, of the earth itself' (ibid. VI, p. 288).

A deep and hard-bottomed lake for Thoreau is symbolic of a kind of highly philosophical self-reflexivity, rather than of merely narcissistic self-contemplation. For him 'a lake is the landscape's most beautiful and expressive feature. It is the earth's eye; looking into which the beholder measures the depth of his own nature' (Thoreau 1982, p. 435).[3] The swamp, by contrast, for Thoreau is shallow and soft, the first birth of nature:

that central meadow and pool in Gowing's Swamp is its very navel, *omphalos*, where the umbilical cord was cut that bound it to creation's womb. Methinks every swamp tends to have or suggests such an interior tender spot. The sphagnous crust that surrounds the pool is pliant and quaking, like the skin or muscles of the abdomen; you seem to be slumping into the very bowels of the swamp. (ibid. p. 524)[4]

The surface of the swamp is the soft spot of nature, even the breasts of Mother Nature when Thoreau refers to 'the soft open sphagnous centre of the swamp' as 'these sphagnous breasts of the swamp – swamp pearls' (Thoreau 1962, X, pp. 38–9). The soft centre of the swamp, the quaking zone, is also related to the human body for Thoreau as 'the part of you that is wettest is fullest of life' (ibid. X, p. 262). Unlike many of his contemporaries, such as Stowe, who figured slavery as a swamp, slavery for Thoreau was the part of the body politic fullest of death: 'slavery . . . has no life. It is only a constant decaying and a death, offensive to all healthy nostrils', dead black waters, unlike the recurring decaying, death and rebirth of living black waters fragrant to Thoreau's healthy sense of smell (ibid VI, p. 365).

For Thoreau 'my temple is the swamp' (ibid. IV, p. 449). He sought refuge and renewal in the swamp as a sacred place, indeed as the inner sanctum, the holy of holies,[5] to which, like the High Priest, he would 'annually go on a pilgrimage' (Thoreau 1993, p. 197). He would perform the ritual of life-giving, self-baptism in the swamp whose waters were not rank poison: 'far from being poisoned in the strong water of the swamp, it is a sort of baptism for which I had waited' (Thoreau 1962, IX. pp. 376–7). Thoreau upset the conventional view that swamps were poisonous by parodying it in his reference to the 'rank and venomous luxuriance in this swamp' (ibid IX, p. 60).

Rather than a reason for avoiding the black swamp of depression and melancholia, Thoreau suggested that 'if you are afflicted with melancholy . . . , go to the swamp' (ibid. X, p. 150). He did not subscribe to a miasmatic theory of malaria nor melancholy. Even at the worst of times he could prescribe a swamp cure: 'when life looks sandy and barren, is reduced to its lowest terms, we have no appetite, and it has no flavour, then let me visit such a swamp as this, deep and impenetrable, where the earth quakes for a rod around you at every step, with its open water where swallows skim and twitter . . .' (ibid. IV, p. 231). When desire is diminished and life is dissatisfying, both represented here orally, the quaking zone of the swamp has depths and softness which the shallowness of its waters belie, and of which the depths of the lake cannot dream. Thoreau values precisely those usual pejorative connotations which attach usually to the 'depth', or horizontal extension, and impenetrability of the swamp.

For Thoreau, the swamp is 'the strength, the marrow of Nature' (Thoreau 1982, p. 613). The strength of nature, for him, lies not in the hard bones of the dry land, but in the soft marrow of the wetlands, what he also called the liquor of nature which feeds the body environmental: 'the very sight of this half-stagnant pond-hole, drying up and leaving bare mud . . . is agreeable and encouraging to behold, as it if contained the seeds of life, the liquor rather, boiled down. The foulest water will bubble purely. They speak to our blood, even these stagnant, slimy pools' (Thoreau 1962, IV, p. 102). They speak to our blood because they contain water which for Thoreau is 'the most living part of nature. This is the blood of the earth' (ibid. XIII, p. 163).

Thoreau's blood circulates with the blood of the earth and with the liquor and marrow of swamps in the body of the earth: 'surely one may as profitably be soaked in the juices of a swamp for one day as pick his way dryshod over sand' (Thoreau 1982, p. 187). Thoreau would prefer the problems of travel through the wetland, the marrow of the earth, than the ease of passage over the dry land, the bones. Yet the problems of travel across the wetland are seasonal anyway in the higher latitudes as 'the deep, impenetrable marsh, where the heron waded and bittern squatted [in summer], is made pervious [in winter] to our swift shoes, as if a thousand railroads had been made into it' (ibid. p. 71). Thoreau sees himself as part of nature, as circulating in the body of nature not via the circulatory system of rivers, but in the stagnant system of marrow through immersion in the swamp by a kind of secular baptism.

Without the wetland, the world would fall apart. The wetland feeds and hold together the skeleton of the body of nature. Without the wetland there would be nothing to replenish the skeletal system of the dry land, the backbones of mountain ranges, the ribs of ridges, the limbs of peninsulas and capes, and the fingers of land reaching into the sea all of which (including the marrow of the wetlands) supply and make possible the fertile plains, prairies and steppes on which agriculture takes place, on which industry depends, on which cities 'live', or more precisely which they parasitically suck dry.

Instead of the standard rhetoric of swampspeak in which the swamp is a place of death and disease, for Thoreau the swamp is the stuff of life and death. Indeed, for Thoreau, 'death is only the phenomenon of the individual or class. Nature does not recognise it. . . Death' is 'a law and not an accident – It is as common as life . . . the law of their [flowers'] death is the law of new life' (Richardson 1986, pp. 114, 115). The swamp, as with nature generally, upsets the hard and fast distinction between life and death. Thoreau inverts the morbid Christian orthodoxy of the saying from the Book of Common Prayer that 'in the midst of life we are in death' by maintaining that 'in the midst of death we are in life' (Thoreau 1993, pp. 100–1; 1962, XIV, p. 109). In the midst of death in the swamp we are in life. The swamp as marrow is constantly being renewed by the life-blood of the earth and constantly renews the bones of the body of the earth.

One of the attractions of the swamp for Thoreau, especially in winter, was that here was a place on which no other 'man' had left a trace, and so it was a place where Thoreau could leave his mark on a *tabula rasa*: 'I love to wade and flounder through the swamp now, these bitter cold days when the snow lies deep on the ground . . . to wade through the swamps, all snowed up, untracked by man' (Thoreau 1962, VIII, p. 99). Unlike the snow of field, pond, or road, the snow of the swamp could remain untracked for a time in order to allow Thoreau to write his own message on its clean sheet, its 'blank page' (ibid. VIII, pp. 160, 167) without fear of interruption or interference from fellow humans, especially citizens, those denizens of the city.

After wading around in a swamp Thoreau felt like an explorer: 'I seemed to have reached a new world, so wild a place . . . far away from human society. What's the need of visiting far-off mountains and bogs, if a half-hour's walk will carry me into such wildness and novelty' (ibid. IX, p. 42). Thoreau explored swamps not just physically but also metaphysically. Indeed, he did not even need to go on a half-hour's walk visiting bogs to be carried into wildness: 'it is in vain to dream of wildness distant from ourselves. There is none such. It is the bogs in our brain and bowels, the primitive vigor of Nature in us, that inspires that dream. I shall never find in the wilds of Labrador any greater wildness than in some recess in Concord, that is than I import into it' (ibid. IX, p. 43). Wild(er)ness is a cognitive, corporeal and cultural experience, not a geographical category of (wet)land conservation or use, or lack of it indigenous or industrial.

Thoreau saw the swamp explorer as a kind of Columbus of the new world of swamps not only without but also within. He asked rhetorically 'is not our own interior white on the chart? black though it may prove, like the coast, when discovered' (Thoreau 1982, p. 560). He then exhorts his readers also to 'be a Columbus to whole new continents and worlds within you, opening new channels, not of trade, but of thought' (ibid.). The interior is either a kind of swamp in winter, a frozen *tabula rasa*, to be explored, mapped, written upon and so colonised or a swamp in summer with its quaking surface which could be decolonised and demapped.

For Thoreau it is the screech owls, or more precisely 'their dismal scream', which best express his view of wetlands as dialogic other:

I love to hear their wailing . . . as if it were the dark and tearful side of music . . . They are the spirits, the low spirits and melancholy forebodings, of fallen souls that once in human shape night-walked the earth and did the deeds of darkness, now expiating their sins with their wailing hymns or threnodies in the scenery of their transgressions. They give me a new sense of the variety and capacity of that nature which is our common dwelling. (ibid. p. 376)

Nature has just as much capacity for evil as it, or 'she,' has for good. Nature is not all lightness and goodness for Thoreau but also has its dark and evil side. Yet the owls are unlike the stymphalian birds in that they are not a monstrous deviation from nature which define and maintain the norm by contrast, but are a part of nature.

Nature for Thoreau is both 'our common dwelling', our home, our domestic setting of steam rising from kettle and swamp, and 'this vast, savage, howling mother of ours' from whose breast 'we are so early weaned . . . to society' (ibid. p. 621). Nature for Thoreau, unlike for his contemporaries and for western 'man' in general, is both homely and unhomely, canny and uncanny. It is both a place of goodness and light perhaps exemplified by the clear 'eyes' of the lake and pond, and a place of life and death, light and dark represented by the 'marrow' of the swamp. Thoreau's double vision, arguably postmodern *avant la lettre*, embraces and entertains both at once without any sense of contradiction between them. The swamp is not a place of melancholy and madness for Thoreau, but a place where melancholy and madness are mediated and alleviated.

The screech owls function for Thoreau as a kind of post-christian scapegoat which instead of being driven off into the pre-modern wilderness to bear the sins of 'men' away from civilisation and the city, are part and parcel of the postmodern wilderness, in which 'men' can find the sacred and solace, can find refuge and sanctuary from the rigours and stresses of modern city life:

I rejoice that there are owls. Let them do the idiotic and maniacal hooting for men. It is a sound admirably suited to swamps and twilight woods which no day illustrates, suggesting a vast and undeveloped nature which men have not recognised. They represent the stark twilight and unsatisfied thoughts which all have. All day the sun has shone on the surface of some savage swamp . . . but now a dismal and fitting day dawns, and a different race of creatures awakes to express the meaning of Nature there. (ibid. p. 377)

The owls suggest a pre-modern wetland which has not yet been subject to a patriarchal and capitalist developmental and industrial technological imperative, yet which is now subjected to that imperative in the very act of naming it as 'vast and undeveloped' with its meanings expressed by owls.

The postmodern wetland is worlds away from the melancholic marshes and the Slough of Despond: 'there can be no very black melancholy to him who lives in the midst of Nature and has his senses still' (ibid. p. 382). The place par excellence in which to live literally in the midst of nature, even up to one's chin, is the swamp. Given the difficulties the swamp poses for travel, it is the perfect place to still the senses, and the limbs, and allow the swamp to write on the body, not as a *tabula rasa*, but as a responsive surface. As for dwellings, Thoreau enjoins us to 'bring your sills up to the very edge of the swamp, then (though it may not be the best place for a dry cellar)' (ibid. p. 612). The slimy edge of the swamp for Thoreau is not the place from which to flee for the bright and sublimed city lights, but the place to live for the bright swamp lights of ignited marsh gases which do not lead to madness, but could even lead to Thoreau's ultimate goal: 'unto a life which I call natural I would gladly follow even a will-o'-the-wisp through bogs and sloughs unimaginable, but no man nor firefly has shown me the causeway to it' (ibid. p. 625).

Thoreau seems to be developing a conservation language which would counter the standard Romantic perception that 'unless Nature sympathises with and speaks to us, as it were, the most fertile and blooming regions are barren and dreary', in other words, are a modern wasteland. (Thoreau 1962, X, p. 252). The postmodern wetland, by contrast, is where nature does not necessarily sympathise with us, nor we with it, but speaks to us, as the screech owls do, in the most fertile and blooming regions of the swamp. The swamp may be bare, but certainly not barren: 'in swamps where there is only here and there an evergreen tree amid the quaking moss and cranberry beds, the bareness does not suggest poverty' (Thoreau 1982, p. 195). The bareness does not suggest barrenness but fertility. Swamp water is living.

The postmodern wetland may not be picturesque in the conventional sense of possessing appropriate qualities of form, texture, depth of field and point of view. Perhaps that is why it has been regarded as barren and dreary. If the wetland had been regarded as picturesque, perhaps its perceived uselessness would not have been held so strongly against it. Perhaps if the wetland could now be regarded as or some of its objects as beautiful, the fact that it is 'useless' as its stands for agriculture or urban development would not matter so much. For Thoreau, 'whatever we have perceived to be in the slightest degree beautiful is of infinitely more value to us than what we have only as yet discovered to be useful and to serve our purpose' (Thoreau 1993, p. 144). The trouble with wetlands is that they have been regarded as lacking both beauty and utility.

The swamp may lack the typical characteristics of beauty, but it does possess gradation which Thoreau saw as one of the fundamental aesthetic and ecological hallmarks of nature: 'nature loves gradation . . . the swamp was variously shaded, or painted even, like a rug, with the sober colours running gradually into each other' (Richardson 1986, p. 360). Rather than subjecting wetlands to an aesthetic and utilitarian, even capitalist, imperative, perhaps it would be preferable to see wetlands as fulfilling vital, ecological functions necessary for life on earth to be sustained. Nature not only loves gradation in colour, but also gradation between land and water, life and death, light and darkness.

SIDNEY LANIER AND THE ALCHEMY OF WETLANDS

Lanier, like Thoreau his near contemporary, turned the rhetoric of standard swampspeak against itself by displacing the usual pejorative associations from the swamp to the city. Lanier, like Thoreau, saw modern industrial capitalism and modern city life as a swamp, or at least giving rise to the malaria (in the literal sense of 'bad air') and malaise associated with swamps. Lanier referred in no uncertain terms to 'that universal killing ague of modern life – the fever of the unrest of trade. . .' (Lanier (1875) 1973, p. 13). As an antidote to the stresses of modern capitalist life Lanier prescribes life in Florida with 'the smooth and glittering suavities of its lakes . . .', and its swamps, and its rivers, including the lake-like and swampy Ocklawaha River, 'a river without banks' (ibid. pp. 13, 24).

Unlike the swamp in standard swampspeak which emits miasma and mephitic vapours, for Lanier 'the glassy dream of a forest over which we sailed appeared to send up exhalations of balms and odors and stimulant pungencies' (ibid. p. 21). These exhalations give rise not only to a pleasant bouquet but also to a counter-rhetoric, 'to the floating suggestions of the unutterable that come up, that sink down, that waver and sway hither and thither' (ibid. p. 23). From the 'depths' or shallows of the wetland the unutterable comes welling up to defy not only standard swampspeak but language itself.

The unutterable is not the sublime about which much can and has been written, but the slime about which little has and can be written because it is slippery and cannot be grasped or defined, even as a place of death, or life, or life and death. The swamp for Lanier is not a place of death and disease, but of life and death: 'it is endless creation succeeded by endless oblivion' (ibid. p. 29). Lanier addresses the marshes of Glynn in Georgia as 'Reverend Marsh, low-crouched along the sea' as if giving birth and as 'Old chemist, rapt in alchemy, . . . The menstruum that dissolves all matter' (Lanier 1969, p. 63). The marsh, indeed the wetland in general, performs the alchemical transformation of base, or dead, or better solid matter into the liquid gold of new life (not only the black gold of swamp power, or 'gas').

This menstrual process of dissolving solid matter into liquid fluid is not the sublimation of the physical into the metaphysical, nor of the sexual into the philosophical, nor of the swampy into the intellectual though it can participate in the transformation of solids into gas. Nor does this process involve the application of the heat and noise of those thermodynamic technologies of modern capitalist industry, of what Lanier, like Stowe, calls 'our tempting hot modern civilisation' (Lanier 1973, p. 14).

Yet as the term 'menstruum' implies, the alchemical process of the wetland is figured as the ripening of the seed within the womb fed by the placenta, so the reverend marsh is more precisely a reverend marsh mother rather than a reverend father. And the reverend mother gives birth by the sea crouched in a 'natural' child-birth position in that most fertile of all regions, the intertidal zone. Rather than a horrifying hell of nightmares of birth, for Lanier 'the marsh with flooded streams/Glimmers, a limpid labyrinth of dreams' which is 'like a lane into heaven that leads from a dream' (Lanier 1969, pp. 64, 47).

JOHN (MUIR) OF THE SWAMPS

Swamps played an important part in the life of John Muir, 'the father of our [United States'] national parks and forest reservations' (Tilden 1970, p. 537), or at least of the idea of them (Badè 1924, II, pp. 392, 445). Better known as 'John of the Mountains', a case could be made for renaming him 'John of the Swamps' for the formative part swamps played in his life (Wolfe 1979). Indeed, Muir's experience in a Canadian swamp in 1864 constituted a kind of conversion experience to the new religion of what Rupert Sheldrake calls 'panentheism'.[6] Muir wrote of this swamp that 'land and water, life and death, beauty and deformity, seemed here to have disputed empire and all shared equally at last' (Badè 1924, I, p. 118). The swamp disputed imperial attempts to classify it as land or water, to mark it on maps, to construct the straight lines of drain, railway and road across and through it, and to vilify it as a place of death and deformity; in short, to destroy it by subjecting it to an imperial imperative which must needs own everything exclusively. As Thoreau said, 'Nature abhors a straight line' (Thoreau 1962, IX, p. 281), but empire adores it.

The idea of sharing anything equally, especially with something as useless and ugly as a swamp, is completely inimical to empire. And the idea of life and death being shared equally in the swamp's ecology is totally anathema to the death-drive of empire. Empire, not the wetland, is the orally sadistic ravening maw which acquisitively consumes everything in its path; the wetland, not the empire, is a generous and reciprocating giver of life. Empire, not the wetland, is the grim reaper, the harbinger of death.

Beside refusing the logic of the imperial death-drive, Muir also refused the either/or logic of the swamp as either land or water, place of life or death, aesthetically beautiful or ugly. For him it was a place of both/and, of both land and water, life and death, beauty and ugliness. Muir bemoaned the fact that 'on no subject are ideas more warped and pitiable than on death' (Badè 1924, I. p. 164; Muir (1916) 1992, p. 70). Against what he called 'morbid death orthodoxy' he celebrated 'the friendly union of life and death so apparent in Nature', not least, and perhaps most powerfully, to be found in the wetland (Badè 1924, I, pp. 164–5; Muir (1916) 1992, p. 70). In nature for Muir there was to be found 'the beautiful blendings and communions of death and life, their joyous inseparable unity' such as he found in the swamps of Canada (Badè 1924, I. p. 165; Muir (1916) 1992, p. 70).

If one feature could be said to characterise the difference between standard swampspeak and the counter-tradition it would be the emphasis in the latter on the wetland as a place of both life and death, whereas for the former it is only a place of death. As Muir's first biographer says, 'in the eyes of other people the spirit of death seemed to brood over these swamps, but to him they were reservoirs of life' (Wolfe 1978, p. 93). Indeed, for Muir, 'everything in Nature called destruction must be creation . . . everything is flowing' and 'there are no harsh dividing lines in nature', not between life and death, nor between land and water (Muir (1911) 1987, pp. 229, 230). Like Thoreau, Muir believed that nature loves gradation not just in colour nor only between land and water but even between life and death.

ALDO LEOPOLD AND A WETLANDS ALMANAC

Leopold is far better known as the writer of *A Sand County Almanac*, the title of his most famous book and the conservationist's bible, than he is for writing a word for wetlands. Yet, as elsewhere in his writings, there are sprinkled throughout the *Almanac* references to and celebrations of wetlands, even elegies for their loss, such as in its 1947 'Foreword' not published until forty years later in which Leopold relates how:

my first doubt about man in the role of conqueror arose when I was still in college. I came home over Christmas to find that land promoters, with the help of the Corps of Engineers, had dyked and drained my boyhood hunting grounds on the Mississippi River bottoms. The job was so complete that I could not even trace the outlines of my beloved lakes and sloughs under their new blanket of cornstalks. (Leopold 1987, p. 282)

Leopold bemoaned, even mourned, the loss of his slough of respond as the death of an old boyhood friend killed by modern engineering. Similarly and recently David Suzuki has described how 'my beloved swamp' where he used to play as a boy 'is covered by an immense shopping mall and parking lot', those icons of late capitalist postmodernity (Suzuki 1991, pp. vi–vii).

Leopold's most powerful elegiac writing about wetlands comes in one of his 'Sketches Here and There' entitled, aptly enough, 'Marshland Elegy' which celebrates the life and death of sandhill cranes.[7] For Leopold marshes may be melancholy places not because of any despondency associated with sinking into them but because 'the sadness discernible in some marshes arises, perhaps, from their once having harboured cranes' (Leopold 1949, p. 97). Mourning for the loss of cranes, not for the loss of the ego, is experienced by him as the melancholic marshes. Cranes have been killed off by drainage when 'the marsh was gridironed with drainage canals' (ibid. p. 100). Gridironing is a very appropriate metaphor for the straight canals, ditches and drains which are the devices of empire that colonise nature.

It is also an apt description for the way in which the marsh became a kind of 'level playing field' for the farmer and the killing fields of the crane. Leopold relates how 'for them, the song of the power shovel came near being an elegy. The high priests of progress knew nothing of cranes, and cared less. What is a species more or less among engineers? What good is an undrained marsh anyhow?' (ibid. p. 100). This slogan could well be the motto of imperialist drainage, the 'highest' drainage of capitalism to modify Lenin, scrawled on the side of the monstrous power shovel or dredger like a war machine, the kind of war fascists loved to fight, and win perhaps because it was so easy to win. It is the power shovel for Leopold, as it is the dredger and steam shovel in Davison's and Swift's novels, that is the marsh monster, not a mythical Grendel-type figure.

Farm and marsh, it seems, cannot co-exist in modern industrial agriculture, certainly not in mutually beneficial symbiosis: 'man cannot live by marsh alone, therefore he must needs live marshless. Progress cannot abide that farmland and marshland, wild and tame, exist in mutual toleration and harmony' (ibid. p. 162). 'Man' cannot live by marsh alone, but then 'he' cannot live by bread alone either, nor can 'he' live without marsh and bread therefore 'he' must live with farm and marsh; he cannot live marshless or farmless, therefore 'he' must live marshful and farmful in mutual toleration, or more preferably mutual aid to use Peter Kropotkin's term (Kropotkin 1991). Otherwise, as Leopold puts it, 'blue lake becomes green bog, green bog becomes caked mud, caked mud becomes a wheatfield' (Leopold. p. 162) and so in turn wheatfield becomes a suburb, a shopping mall, a parking lot, a wasteland, a postmodern late capitalist waste land of sterile concrete and scorching bitumen.

Leopold was as scathing of conservationists, or at least of what he called 'alphabetical conservationists', when it comes to marshes as he is of the drivers and backers of monstrous dredgers and shovels: 'a roadless marsh is seemingly as worthless to the alphabetical conservationist as an undrained one was to the empire builders. Solitude, the one natural resource still unendowed of alphabets, is so far unrecognised as valuable only by ornithologists and cranes' (ibid. p. 101). An alphabetical conservationist, even ornithological 'twitcher', is one who ticks off his or her sighting of a sandhill crane or spoonbill in his or her field guide then moves on to find the next species oblivious to the crane's or grebe's wetland habitat, not being able to see the wood for the trees (ibid. p. 160).

Leopold, the educator for much of his life, asked 'is education possibly a process of trading awareness for things of lesser worth?' (ibid. p. 18). Leopold feared that education is 'learning to see one thing by going blind to another. One thing most of us have gone blind to is the quality of marshes' (ibid. p. 158). Yet the quality which Leopold valued most highly in marshes was one which required a kind of willing blindness or at least a suspension of the sense of sight, and not of other senses, particularly of hearing which could be heightened in going to the marsh, especially if one went 'too early': 'to arrive too early in the marsh is an adventure in pure listening; the ear roams at will among the noises of the night, without let or hindrance from hand or eye' (ibid. p. 61). The sense of hearing implies immediacy with the living animal heard unlike the mastery of hand or eye over its object. Yet hearing is not as immediate nor as intimate as the sense of smell.

For Leopold the marsh was a vestige of pre-modern wilderness as 'a sense of time lies thick and heavy on such a place' (ibid. p. 96). Sandhill cranes themselves represented wildness because their symphony of cries was literally 'a blast from the past': 'when we hear his [sic] call we hear no mere bird. We hear the trumpet in the orchestra of evolution. He [sic] is the symbol of our untameable past, of that incredible sweep of millennia which underlies and conditions the daily affair of birds and men' (ibid.). The sandhill crane speaks the wetland as the wilderness foundation of modern society in a language which all could hear if they had ears to hear it, and if their eyes could stop seeing.

Like Thoreau whose fourteen-volume *Journal* he had received as a wedding present from his mother and for whom 'the most alive is the wildest' (Thoreau 1982, p. 611), Leopold argued that 'the ultimate value in these marshes is their wildness, and the crane is wildness incarnate. But all conservation of wilderness is self-defeating, for to cherish we must see and fondle, and when enough have seen and fondled, there is no wilderness left to cherish' (Leopold 1949, p. 101). Leopold generally bemoaned the fact that 'man always kills the thing he loves, and so we the pioneers have killed our wilderness' (ibid. p. 148). Perhaps it is more precisely patriarchal western man who kills the thing he loves, whereas indigenous humankind 'love the things they kill' (Stow 1958, p. 172).[8] The former is a code based on retribution and sadism, the Law of the Father, the latter on generosity and reciprocity, the Body of the Mother.

Leopold, like Thoreau, wrote a word for wetlands, but he also seemed to try to out-Thoreau Thoreau when it came to going into them. If Thoreau loved to stand up to his chin in some retired swamp, Leopold wished he were a muskrat 'eye-deep in the marsh' (Leopold 1949, p. 19). For Leopold, the only way to see the marsh is at its level, not from above surveying it standing up and certainly not from the air or on the map mastering it imperially from above. Leopold even fulfilled his wish when 'one day I buried myself, prone, in the muck of a muskrat house. While my clothes absorbed local colour, my eyes absorbed the lore of the marsh' (ibid. p. 160). And presumably his nose and ears absorbed its smells and sounds too (though there are no references to either) as he was absorbed into the marsh, not in an act of oral sadism on its part nor in an act of sadistic epistemophilia on his part, nor as a simulation of death, but as a celebration of life, and death.

SEAMUS HEANEY AND IRISH BOGS

From the earliest of his published poetry Seamus Heaney has celebrated the centrality of the bog to the Irish landscape, to Irish history and to Irish cultural identity. He even goes so far as to suggest in 'The Tollund Man', one of several poems devoted to 'bog men', that 'I could risk blasphemy,/Consecrate the cauldron bog/Our holy ground . . .' (Heaney 1980, p. 126). Like Thoreau, Heaney sees the wetland as a sacred place. Part of the holiness of the bog consists in its distinctive Irishness, particularly to the Irish landscape. Heaney seeks a sense of Irish cultural, if not national, identity in the bog.

The bog has been associated with death, and with life. Heaney feels kinship with the bog, or at least he feels 'kinned by hieroglyphic/peat' which writes on him as the swamp oak wrote on Thoreau's heart:

Quagmire, swampland, morass:
the slime kingdoms,
domains of the cold-blooded,
of mud pads and dirtied eggs.

But bog
meaning soft,
the fall of windless rain,
pupil of amber.

Ruminant ground,
digestion of mollusc
and seed-pod,
deep pollen bin.

Earth-pantry, bone-vault,
sun-bank, embalmer
of votive goods
and sabred fugitives.

Insatiable bride.
Sword-swallower,
casket, midden,
floe of history.

Ground that will stip
its dark side,
nesting ground,
outback of my mind.
(Heaney 1980, pp. 196–7)

In contrast to Thoreau for whom there is a hard bottom everywhere, for Heaney 'the wet centre is bottomless' and in contrast to
his earlier compatriot poet W. B. Yeats for whom 'the centre
cannot hold,' for Heaney 'this centre holds' despite, or perhaps
because of, its fluidity (ibid. pp. 86, 198). Thoreau was looking for
some sort of bedrock of reality, whereas for Heaney there is none;
Yeats was looking for some fixed reference point in the flux of
modernity whereas for Heaney the only centre which holds in the
swamp of postmodernity is one that changes like the bog.

Heaney is fascinated by what he calls 'a mothering smell of
wet', though perhaps not emanating from a bog (Heaney 1979,
p. 53). The bog is 'a slimy birth-cord' which would attach him too
strongly to the spawn of mother earth, or specifically to the 'Bog
Queen' of Ireland herself, or which would drag him down into
'the black maw/Of the peat', into its orally sadistic *vagina dentata*
(Heaney 1980, pp. 186, 189). After being immured in the bog and

masticated by peat's teeth the interior journey continues on into the fen of the digestive system with 'those dark juices working' where, as for the Bog Queen, 'the seeps of winter/digested me' (ibid. p. 187).

For Heaney the bog and its processes of decay and digestion, death and life, are 'the vowel of the earth' rather than the bowels of the earth. The consonants of the earth are the hard, dry land, the harsh, sibilant sounds (like the souls of the sullen stuck in the slime of the Styx), the bones of the land, the exteriority of the Law of the Father, whereas the vowels of the earth are the wetlands, the round, open sounds, the marrow, the breasts, the womb, the placenta, the interiority of the Body of the Mother, the grotesque lower bodily and earthly stratum, the place of life and death – living black waters. Please help save them.

NOTES

1. Both these books have chapters devoted to Thoreau, Muir and Leopold but no mention is made of the importance of marshes and swamps in their work.
2. See, for example, Michaels 1977, pp. 132–49.
3. See also pp. 67, 339, 437 and 527.
4. See also Thoreau 1962, IX, p. 394. Carl Bode coyly excludes the first sentence and last phrase when he quotes from this passage in his 'Epilogue by the Editor', p. 686.
5. See the epigraph to this book taken from Thoreau 1982, p. 613.
6. For a discussion of this episode in Muir's life, see Turner 1985, pp. 113–16; and Fox 1981, pp. 43–4. For Muir's life and work see also Cohen 1984. For 'panentheism' see Sheldrake 1991, p. 198. I am grateful to Ron Blaber for drawing my attention to the latter.
7. For a discussion of the role of sandhill cranes in Leopold's life, see Meine 1988, pp. 329–31.
8. See also Giblett 1993, pp. 541–59.

Bibliography

Abraham, N. and M. Torok (1984) 'A poetics of psychoanalysis: "The lost object – me"', *SubStance*, 43, pp. 3–18.

Allen, P. (1947) 'Etiological theory in America prior to the civil war', *Journal of the History of Medicine and Allied Sciences*, 2, pp. 489–520.

Appleyard, R. T. and T. Manford (eds) (1979) *The Beginning: Early European Discovery and Settlement of Swan River Western Australia*, University of Western Australia Press.

Aristotle (1927) 'Problemata' (trans. E. S. Forster), *The Works*, VII (ed. W. D. Ross), Oxford: Clarendon Press.

Aristotle (1930) 'De generatione et corruptione' (trans. H. H. Joachim), *The Works*, II (ed. W. D. Ross), Oxford: Clarendon Press.

Atwood, M. (1991) 'The bog man', *Wilderness Tips*, London: Bloomsbury.

Bachelard, G. (1983) *Water and Dreams: An Essay on the Imagination of Matter* (trans. E. R. Farrell), The Pegasus Foundation/The Dallas Institute of Humanities and Culture.

Bachofen, J. J. (1967) *Myth, Religion and Mother Right: Selected Writings* (trans. R. Manheim), Princeton University Press.

Badè W. F. (1924) *The Life and Letters of John Muir*, I and II, Boston: Houghton Mifflin.

Bakhtin, M. (1984) *Rabelais and His World* (trans. H. Iswolsky), Indiana University Press.

Ballard, J. G. (1983) *The Drowned World* (1963), London: Dent.

Bass, R. D. (1959) *Swamp Fox: The Life and Campaigns of General Francis Marion*, London: Alvin Redman.

Battye, J. S. (ed.) (1912) *The Cyclopedia of Western Australia*, Vol. I, *An Historical and Commercial Review, Descriptive and Biographical Facts, Figures and Illustrations, An Epitome of Progress*, Perth: Cyclopedia.

Baudrillard, J. (1990) *Seduction* (trans. B. Singer), London: Macmillan.

Bauman, Z. (1973) *Culture as Praxis*, London: Routledge and Kegan Paul.

Bekle, H. (1981) 'The wetlands lost: drainage of the Perth lake systems', *Western Geographer*, 5, 1–2, pp. 22–41.

248

Bekle, H. and J. Gentilli (1993) 'History of the Perth lakes', *Early Days*, 10, 5, pp. 422–60.

Bell, V. (1981) *Swamp Water* (1941), University of Georgia Press.

Bell, V. Jr (1981) 'Foreword', in V. Bell, *Swamp Water* (1941), repr. University of Georgia Press.

Belloc, H. (1923) *The Road*, Manchester: Charles W. Hobson.

Belloc, H. (1970) *Hills and the Sea* (1906), Westport, Connecticut: Greenwood Press.

Benjamin, J. and A. Rabinbach (1989) 'Foreword', in K. Theweleit, *Male Fantasies*, II: *Male Bodies: Psychoanalyzing the White Terror* (trans. C. Turner and E. Carter), Cambridge: Polity.

Beowulf and the Finnesburg Fragment (trans. J. R. Clark Hall), new edition, London: George Allen and Unwin.

Berman, M. (1983) *All That Is Solid Melts Into Air: The Experience of Modernity*, London: Verso.

Bermingham, A. (1986) *Landscape and Ideology: The English Rustic Tradition 1740–1860*, University of California Press.

Bermingham, A. (1990) 'Reading Constable', in *Reading Landscape: Country – City – Capital* (ed. S. Pugh), Manchester University Press.

Bodard, L. (1967) *The Quicksand War: Prelude to Vietnam* (trans. P. O'Brian), London: Faber and Faber.

Bode, C. (1982) 'Epilogue by the editor', in *The Portable Thoreau* (ed. C. Bode), New York: Penguin.

Bold, W. E. (1939) 'Perth – the first hundred years: the story of the municipal development of our city', *'Early Days': Journal and Proceedings of the Western Australian Historical Society*, 3, 2, pp. 29–41.

Bolton, G. (1989) 'Perth: A Foundling City', *The Origins of Australia's Capital Cities* (ed. P. Statham), Cambridge University Press.

Bolton, G. and S. Hunt (1978) 'Cleansing the dunghill: water supply and sanitation in Perth 1878–1912', *Studies in Western Australian History*, 2, pp. 1–17.

Bonaparte, M. (1949) *The Life and Works of Edgar Allan Poe: A Psycho-Analytic Interpretation* (trans. J. Rodker), London: Imago.

Bonaparte, M. (1953) *Female Sexuality*, London: Imago.

Boyle, T. C. (1986) 'Greasy Lake', *Greasy Lake and Other Stories*, Harmondsworth: Penguin.

Browne, M. W. (1965) *The New Face of War*, Indianapolis: Bobbs-Merrill.

Bruce-Chwatt, L. J. (1976) 'Ague as malaria (an essay on the history of two medical terms)', *The Journal of Tropical Medicine and Hygiene*, 79, pp. 168–76.

Bruce-Chwatt, L. J. (1988) 'History of malaria from prehistory to eradication', *Malaria: Principles and Practice of Malariology* (ed. W. H. Wernsdorfer and I. McGregor), Edinburgh: Churchill Livingstone.

Buker, G. E. (1975) *Swamp Sailors: Riverine Warfare in the Everglades 1835–1842*, The University Presses of Florida.

Burchett, W. G. (1965) *Vietnam: Inside Story of a Guerilla War*, New York: International Publishers.

Burke, E. (1958) *A Philosophical Enquiry Into the Origin of Our Ideas of the Sublime and Beautiful* (ed. J. T. Boulton), London: Routledge and Kegan Paul.

Burton, R. (1932) *The Anatomy of Melancholy* (ed. Holbrook Jackson), London: Dent.

Buttinger, J. (1958) *The Smaller Dragon: A Political History of Vietnam*, New York: Praeger.

Buttinger, J. (1967a) *Vietnam: A Dragon Embattled*, I, *From Colonialism to the Vietminh*, New York: Praeger.

Buttinger J. (1967b) *Vietnam: A Dragon Embattled*, II, *Vietnam at War*, New York: Praeger.

Buttinger, J. (1972) *A Dragon Defiant: A Short History of Vietnam*, New York: Praeger.

Byrd, W. (1966) *The Prose Works* (ed. Louis B. Wright), Harvard University Press.

Carrouges, M. (1954) *Les machines célibataires*, Paris: Arcanes.

Carter, P. (1987) *The Road to Botany Bay: An Essay in Spatial History*, London: Faber.

Certeau, M. de (1984) *The Practice of Everyday Life* (trans. S. F. Rendall), Berkeley: University of California Press.

Certeau, M. de (1986) 'The arts of dying: celibatory machines', *Heterologies: Discourse on the Other* (trans. B. Massumi), Manchester University Press.

Chadwick, E. (1965) *Report on the Sanitary Condition of the Labouring Population of Great Britain* (ed. M.W. Flinn) (1842), Edinburgh University Press.

Chamberlain, M. (1975) *Fenwomen: A Portrait of Women in an English Village*, London: Virago.

Cicero (1972) *The Nature of the Gods* (trans. H. C. P. McGregor), London: Penguin.

Cixous, H. and C. Clément, *The Newly Born Woman* (trans. B. Wing), Manchester University Press.

Clausewitz, General C. von (1968) *On War*, II and III (trans. Colonel J. J. Graham), new and revised edition by Colonel F. N. Maude, London: Routledge and Kegan Paul.

Cohen, M. P. (1984) *The Pathless Way: John Muir and American Wilderness*, University of Wisconsin Press.

Colebatch Sir H. (ed.) (1929) *A Story of A Hundred Years*, Perth: Fred W. Simpson.

Coles, B. and J. Coles (1989) *People of the Wetlands: Bogs, Bodies and Lake-Dwellers*, London: Thames and Hudson.

Collins, M. L. and C. Pierce (1976) 'Holes and slime: sexism in Sartre's psychoanalysis', *Women and Philosophy* (ed. C. Gould and M. W. Wartofsky), New York: Putnam.

Collins, S. H. (1829) *The Emigrants Guide to the United States of America*, Hull: Joseph Noble.

Collins, W. (1974) *The Woman in White* (1859–60), Harmondsworth: Penguin.

Conrad, J. (1977) 'The lagoon', *Tales of Unrest* (1898), Harmondsworth: Penguin.

Corbin, A. (1994) *The Foul and the Fragrant: Odour and the Social Imagination*, London: Picador.

Covell, G. (1967) 'The story of malaria', *The Journal of Tropical Medicine and Hygiene*, 70, pp. 281–5.

Creed, B. (1986) 'Horror and the monstrous-feminine: an imaginary abjection', *Screen*, 27, 1, pp. 44–70.

Crosby, A. W. (1986) *Ecological Imperialism: The Biological Expansion of Europe, 900–1900*, Cambridge University Press.

Currie, W. (1799) 'An enquiry into the causes of the insalubrity of flat and marshy situations; and directions for preventing or correcting the effects thereof', *Transactions of the American Philosophical Society*, 4, pp. 127–42.

Curtin, P. D. (1964) *The Image of Africa: British Ideas and Action, 1780–1850*, The University of Wisconsin Press.

Daly, M. (1979) *Gyn/Ecology: The Metaethics of Radical Feminism*, London: Women's Press.

Darby, H. C. (1956) *The Draining of the Fens*, 2nd edn, Cambridge University Press.

Darby, H. C. (1974) *The Medieval Fenland*, 2nd edn, Newton Abbot: David and Charles.

Darby, H. C. (1983) *The Changing Fenland*, Cambridge University Press.

Davis, H. (1962) *The Great Dismal Swamp: Its History, Folklore and Science*, revised edition, no publisher.

Davison, L. (1993) *Soundings*, University of Queensland Press.

Deleuze, G. (1984) *Kant's Critical Philosophy: The Doctrine of the Faculties* (trans. H. Tomlinson and Habberjam), London: Athlone.

Deleuze, G. (1989) 'Coldness and cruelty', *Masochism*, New York: Zone.

Deleuze, G. and F. Guattari (1977) *Anti-Oedipus: Capitalism and Schizophrenia* (trans. R. Hurley, M Seem and H. R. Lane), New York: Viking.

Deleuze, G. and F. Guattari (1986) *Nomadology: The War Machine* (trans. B. Massumi), New York: Semiotext(e).

Derrida, J. (1976) *Of Grammatology* (trans. G. Chakravorty Spivak), The John Hopkins University Press.

Derrida, J. (1978) 'Freud and the scene of writing', *Writing and Difference* (trans. A. Bass), London: Routledge and Kegan Paul.

Derrida, J. (1981) *Dissemination* (trans. B. Johnson), London: Athlone.

Derrida, J. (1982) 'White mythology: metaphor in the text of philosophy', *Margins of Philosophy* (trans. A. Bass), Brighton: Harvester.

Dickens, C. (1953) *Martin Chuzzlewit* (1884), London: Collins.

Dickens, C. (1987) *The Mudfog Papers* (1880), Gloucester: Alan Sutton.

Dluhosch, E. (1969) 'Translator's note on the founding and development of St. Petersburg', in I. A. Egorov, *The Architectural Planning of St Petersburg*, Ohio University Press.

Dobson, M. (1980) '"Marsh fever" – the geography of malaria in England', *Journal of Historical Geography*, 6, 4, pp. 357–89.

Douglas, J. D. (ed.) (1962) *The New Bible Dictionary*, Inter-Varsity Fellowship.

Douglas, M. (1966) *Purity and Danger: An Analysis of the Concepts of Pollution and Taboo*, London: Routledge.

Duiker, W. J. (1983) *Vietnam: Nation in Revolution*, Boulder: Westview.

Durand, G. (1969) *Les structures anthropologiques de l'imaginaire: introduction à l'archétypologie générale*, Paris: Bordas.

Duras, M. (1985) *The Lover* (trans. B. Bray), London: Collins.

Duras, M. (1986) *The Sea Wall*, London: Faber.

Durgnat, R. (1975) *Jean Renoir*, London: Studio Vista.

Eagleton, T. (1986) *William Shakespeare*, Oxford: Basil Blackwell.

Eagleton, T. (1990) *The Ideology of the Aesthetic*, Oxford: Basil Blackwell.

Eagleton, T. (1991) *Ideology: An Introduction*, London: Verso.

Earll, R. (1991) 'Shorelines and estuaries: turning the tide', in J. Porritt (ed.), *Save the Earth*, North Ryde: Angus and Robertson.

Easthope, A. (1986) *What A Man's Gotta Do: The Masculine Myth in Popular Culture*, London: Paladin.

Ehrenreich, B. (1987) 'Foreword', in K. Theweleit, *Male Fantasies*, I: *Women, Floods, Bodies, History* (trans. S. Conway), Cambridge: Polity.

Eisler, R. (1987) *The Chalice and the Blade: Our History, Our Future*, San Francisco: Harper and Row.

Eliade, M. (1958) *Patterns in Comparative Religion* (trans. R. Sheed), London: Sheed and Ward.

Engels, F. (1969) 'Preface to the English edition', *The Condition of the Working Class in England* (1882), London: Grafton.

Errington, P. (1957) *Of Men and Marshes*, New York: Macmillan.

Falkiner, S. (1992) *The Writer's Landscape – Wilderness*, East Roseville: Simon and Schuster.

Fall, B. B. (1963) *Street Without Joy: Insurgency in IndoChina, 1946–1963* (3rd rev. edn), London: Pall Mall Press.

Fall, B. B. (1967) *The Two Vietnams: A Political and Military Analysis* (2nd edn), New York: Praeger.

Fanon, F. (1967) *The Wretched of the Earth* (trans. C. Farrington) (1965), Harmondsworth: Penguin.

Ferguson, D. J. (1847) *Colonial Secretaries Office*, vol. 161.

Ferguson, W. (1823) 'On the nature and history of the marsh poison', *Transactions of the Royal Society of Edinburgh*, 9, pp. 273–98.

Flood, J. (1995) *Archaeology of the Dreamtime: The Story of Prehistoric Australia and its People* (rev edn), Sydney: HarperCollins.

Forester, C. S. (1956) *The African Queen* (1935), Harmondsworth: Penguin.

Forster, E. M. (1936) *A Passage to India* (1924), Harmondsworth: Penguin.

Foucault, M. (1970) *The Order of Things: An Archaeology of the Human Sciences*, London: Tavistock.

Foucault, M. (1979) *Discipline and Punish: The Birth of the Prison* (trans. A. Sheridan), Harmondsworth: Penguin.

Fox, S. (no date) *The American Conservation Movement: John Muir and his Legacy* (1981) (repr.), The University of Wisconsin Press.

Frame, J. (1991) 'The Lagoon', *The Lagoon and Other Stories* (1951), London: Bloomsbury.

Freud, S. (1964) *New Introductory Lectures on Psycho-Analysis*, Lecture XXXI, *Standard Edition*, vol. XXII, London: Hogarth.

Freud, S. (1973) *Introductory Lectures on Psychoanalysis*, Pelican Freud Library 1, Harmondsworth: Penguin.

Freud, S. (1979) 'Inhibitions, symptoms and anxiety', *On Psychopathology: Inhibitions, Symptoms and Anxiety and Other Works*, Penguin Freud Library 10, Harmondsworth: Penguin.

Freud, S. (1984) 'The economic problem of masochism', *On Metapsychology: The Theory of Psychoanalysis*, Pelican Freud Library 11, Harmondsworth: Penguin.

Freud, S. (1984) 'Mourning and melancholia', *On Metapsychology: The Theory of Psychoanalysis*, Pelican Freud Library 11, Harmondsworth: Penguin.

Freud, S. (1985) 'Civilization and its discontents', *Civilization, Society and Religion*, Penguin Freud Library 12, Harmondsworth: Penguin.

Freud, S. (1985) '"Civilized" sexual morality and modern nervous illness', *Civilization, Society and Religion*, Penguin Freud Library 12, Harmondsworth: Penguin.

Freud, S. (1985) 'Creative writers and day-dreaming', *Art and Literature: Jensen's Gravida, Leonardo da Vinci and Other Works*, Penguin Freud Library 14, Harmondsworth: Penguin.

Freud, S. (1985) 'The "Uncanny"', *Art and Literature*, Penguin Freud Library 14, Harmondsworth: Penguin.

Fritzell, P. A. (1978) 'American wetlands as cultural symbol: places of wetlands in American culture', in *Wetland Functions and Values: The State of our Understanding* (ed. P. R. Greeson, J. R. Clark and J. E. Clark), Minneapolis: American Water Resources Association.

Gallop, J. (1982) *Feminism and Psychoanalysis: The Daughter's Seduction*, London: Macmillan.

Gare, A. E. (1995) *Postmodernism and the Environmental Crisis*, London: Routledge.

Sir Gawain and the Green Knight (1975) (trans. J. R. R. Tolkien), London: George Allen & Unwin.

Gerson, N. B. (1967) *The Swamp Fox, Francis Marion*, New York: Doubleday.

Giblett, R. (1993) 'Kings in Kimberley watercourses', *Span*, 36, 2, pp. 541–59.

Giblett, R. (1994) 'Terraphagy and the politics of mining', *Proceedings of the Third International Conference on Environmental Issues and Waste Management in Energy and Mineral Production*, Perth, Western Australia: Curtin University of Technology, pp. 55–62.

Gilbert, S. and S. Gubar (1989) *No Man's Land: The Place of the Woman Writer in the Twentieth Century*, II, *Sexchanges*, Yale University Press.

Gimbutas, M. (1982) *The Goddesses and Gods of Old Europe: 6500–3500 BC: Myths and Cult Images*, University of California Press.

Gimbutas, M. (1989) *The Language of the Goddess: Unearthing the Hidden Symbols of Western Civilization*, San Francisco: Harper and Row.

Glacken, C. (1967) *Traces on the Rhodian Shore: Nature and Culture in Western Thought from Ancient Times to the End of the Eighteenth Century*, University of California Press.

Goethe, J. W. (1969) *Faust* (trans. P. Wayne), Harmondsworth: Penguin.

Goethe, J. W. (1970) *Italian Journey* (trans. W. H. Auden and E. Mayer) (1962), Harmondsworth: Penguin.

Graves, R. (1955) *The Greek Myths*, II, Harmondsworth: Penguin.

Green, M. (1979) *Dreams of Adventure, Deeds of Empire*, New York: Basic Books.

Greenblatt, S. (1982) 'Filthy rites', *Daedalus*, 111, 3, pp. 1–16.

Greenwood, M. (1953) 'Miasma and contagion', *Science, Medicine and History: Essays on the Evolution of Scientific Thought and Medical Practice Written in Honour of Charles Singer* (ed. E. Ashworth Underwood), 2, Oxford University Press.

Grey, G. (1841) *Journals of Two Expeditions of Discovery in North-West and Western Australia During the Year 1837, 38 and 39 . . .*, II, London: T. and W. Boone.

Griffin, S. (1978) *Woman and Nature: The Roaring Inside Her*, New York: Harper and Row.

Grosz, E. (1994) *Volatile Bodies: Toward a Corporeal Feminism*, St Leonards, N.S.W.: Allen and Unwin.

Grosz, E. (1995) *Space, Time and Perversion: The Politics of Bodies*, St Leonards, N.S.W.: Allen and Unwin.

Guattari, F. (1989) 'The three ecologies', *New Formations*, 8, pp. 131–47.

Haggard, H. R. (1991) *She* (1887), Oxford University Press.

Haraway, D. (1992), 'The promises of monsters: a regenerative politics for inappropriate/d Others', in *Cultural Studies* (ed. L. Grossberg, C. Nelson and P. A. Treichler), New York: Routledge.

Harley, J. B. (1988) 'Maps, knowledge and power', in *The Iconography of Landscape: Essay on the Symbolic Representation, Design and Use of Past Environments* (ed. D. Cosgrove and S. Daniels), Cambridge University Press, pp. 277–312.

Harley, J. B. (1992) 'Deconstructing the map', in *Writing Worlds: Discourse, Text and Metaphor in the Representation of the Landscape* (ed. T. J. Barnes and J. S. Duncan), London: Routledge, pp. 231–47.

Harris, L. E. (1957) 'Land drainage and reclamation', in *A History of Technology*, III (ed. C. Singer, E. J. Holmyard, A. R. Hall and T. J. Williams), Oxford: Clarendon Press.

Harrison, G. (1978), *Mosquitoes, Malaria and Man: A History of the Hostilities Since 1880*, New York: E. P. Dutton.

Hart, K. (1986) 'Maps of deconstruction', *Meanjin*, 45, 1, pp. 107–16.

Harvey, R. (1991) 'The Sartrean viscous: swamp and source', *SubStance*, 64, pp. 49–66.

Hasluck, A. (1955) *Portrait With Background: A Life of Georgiana Molloy*, Oxford University Press.

Hasluck, P. and F. I. Bray (1930) 'Early mills of Perth', *The Western Australian Historical Society Journal and Proceedings*, 1, 8, pp. 80–1.

Hebdige, D. (1987) 'The impossible object: towards a sociology of the sublime', *New Formations*, 1, pp. 47–76.

Hermand, J. (1980) 'Explosions in the swamp: Jünger's *Worker* (1932)', in *The Technological Imagination: Theories and Fictions* (ed. T. de Lauretis, A. Huyssen and K. Woodward), Madison, Wisconsin: Coda Press.

Hill, C. (1965) *God's Englishman: Oliver Cromwell and the English Revolution*, London: Weidenfeld and Nicolson.

Hippocrates: On Endemic Diseases, Airs, Waters and Places (1969), Arabic Technical and Scientific Texts, 5 (ed. and trans. J. N. Mattock and M. C. Lyons), Cambridge: W. Heffer.

Hippocratic Writings (1978) (ed. G. E. R. Lloyd and trans. J. Chadwick, W. N. Mann, I. M. Lonie and E. T. Withington) (1950), Harmondsworth, Penguin.

Hobsbawm, E. J. (1985) *The Age of Capital 1848–1875*, London: Abacus.

Hobsbawm, E. J. (1995) *Age of Extremes: The Short Twentieth Century 1914–1991*, London: Abacus.

Hochbaum, H. A. (1981) *The Canvasback on a Prairie Marsh* (1944), University of Nebraska Press.

Hoffmann, E. T. W. (1982) 'The Sandman', *Tales* (trans. R. J. Hollingdale), London: Penguin.

Homer (1965) *Odyssey* (trans. R. Lattimore), New York: Harper and Row.

Homer (1967) *The Odyssey* (trans. and ed. A. Cook), New York: W. W. Norton.

Honda, K. (1972) *Vietnam War: A Report Through Asian Eyes*, Tokyo: Mirai-Sha.

Hopkins, G. M. (1953) 'Inversnaid', *Poems and Prose*, Harmondsworth: Penguin.

Huggan, G. (1989) 'Decolonizing the map: post-colonialism, post-structuralism and the cartographic connection', *Ariel*, 20, pp. 115–31.

Hughes, R. (1980) *The Shock of the New: Art and the Century of Change*, British Broadcasting Corporation.

Hunt, S. J. (1980) *Water, The Abiding Challenge* (ed. F. B. Morony), Perth: Metropolitan Water Board.

Hutcheon, L. (1988) *A Poetics of Postmodernism: History, Theory, Fiction*, New York: Routledge.

Huxley, T. H. (1989) 'Struggle for existence', in Peter Kropotkin (ed.), *Mutual Aid: A Factor of Evolution* (1914), Montreal: Black Rose.

Illich, I. (1986) *H_2O and the Waters of Forgetfulness*, London: Marion Boyars.

Irigaray, L. (1977) 'Women's exile', *I and C*, 1, pp. 62–76.

Irigaray, L. (1985) *Speculum of the Other Woman* (trans. G. C. Gill), Cornell University Press.

Irigaray, L. (1985) *This Sex Which Is Not One* (trans. C. Porter), Cornell University Press.

Irigaray, L. (1991) *Marine Lover of Friedrich Nietzsche* (trans. G. C. Gill), Columbia University Press.

Irigaray, L. (1993) *Je, Tu, Nous: Toward a Culture of Difference* (trans. A. Martin), New York: Routledge.

Irigaray, L. (1993) *Sexes and Genealogies* (trans. G. C. Gill), Columbia University Press.

Jackson, J. L. B. (1982) *Not an Idle Man: A Biography of John Septimus Roe, Western Australia's First Surveyor-General, 1797–1878*, West Swan: M. B. Roe.

Jacob, F. (1973) *The Logic of Life: A History of Heredity* (trans. B. E. Spillman), New York: Pantheon.

Jacobson, M. B. (1989) *The Freudian Subject* (trans. C. Porter), London: Macmillan.

Jameson, F. (1984) 'Postmodernism, or the cultural logic of late capitalism', *New Left Review*, 146, pp. 53–92.

Jameson, F. (1984), 'Progress versus utopia; or, can we imagine the future', in *Art After Modernism: Rethinking Representation* (ed. B. Wallis), New York: The New Museum of Contemporary Art.

Jarcho, S. (1970) 'A cartographic and literary study of the word *malaria*', *Journal of the History of Medicine*, 25, pp. 31–9.

Jardine, A. (1985) *Gynesis: Configurations of Woman and Modernity*, Cornell University Press.

Jefferies, R. (1939) *After London and Amaryllis at the Fair* (1885), London: Dent.

Jewett, S. O. (1885) *A Marsh Island*, Cambridge, MA: Houghton, Mifflin.

Journals of Several Expeditions Made in Western Australia During the Years 1829, 1830, 1831 and 1832 . . . (1833/1980), University of Western Australia Press.

Kant, I. (1952) *The Critique of Judgement* (trans. J. C. Meredith), Oxford Clarendon.

Kant, I. (1960) *Observations on the Feeling of the Beautiful and Sublime* (trans. J. T. Goldthwait), University of California Press.

Kelleher, V. (1990) *Brother Night*, London: Julia MacRae.

Kiernan, V. G. (1978) *America: The New Imperialism: From White Settlement to World Hegemony*, London: Zed.

Kingsford, R. (1991) *Australian Waterbirds: A Field Guide*, Kenthurst, N.S.W.: Kangaroo Press.

Kingsley, C. (1877) *His Letters and Memories of his Life*, I (edited by his wife), London: Henry S. King.

Kingsley, C. (1884) 'The Fens', *The Works*, XV, *Prose Idylls*, London: Macmillan.

Kingsley, C. (1908) *Hereward the Wake*, London: Dent.

Kirk, G. S., J. E. Raven and M. Schofield, *The Pre-Socratic Philosophers: A Critical History with a Selection of Texts* (2nd edn), Cambridge University Press.

Klein, M. (1986) 'Early stages of the Oedipus conflict' (1928), *The Selected Melanie Klein* (ed. J. Mitchell), Harmondsworth: Penguin.

Knoepflmacher, U. C. (1973) 'The novel between city and country', *The Victorian City: Images and Realities*, II (ed. H. J. Dyos and M. Wolff), London: Routledge and Kegan Paul.

Kristeva, J. (1982) *Powers of Horror: An Essay on Abjection* (trans. L. S. Roudiez), Columbia University Press.

Kristeva, J. (1986) 'About Chinese women', *The Kristeva Reader* (ed. T. Moi), Oxford: Basil Blackwell.

Kristeva, J. (1989) *Black Sun: Depression and Melancholia* (trans. L. S. Roudiez), Columbia University Press.

Kristeva, J. (1991) *Strangers to Ourselves* (trans. L. S. Roudiez), New York: Harvester Wheatsheaf.

Kropotkin, P. (1991) *Mutual Aid: A Factor in Evolution* (1914), Montreal: Black Rose.

Lacan J. (1977) *Écrits: A Selection* (trans. A. Sheridan), London: Tavistock.

Ladurie, E. Le Roy (1989) 'Introduction', in J. P. Goubert, *The Conquest of Water: The Advent of Health in the Industrial Age* (trans. A. Wilson), Cambridge: Polity Press.

Langley, D. (1982) *Swamp Angel*, Chicago: Academy Edition.

Lanier, S. (1969) *Poems and Letters*, The Johns Hopkins University Press.

Lanier, S. (1973) *Florida: Its Scenery, Climate, and History* (1875), University of Florida Press.

Laplanche, N. and J.-B. Pontalis (1973), 'Sublimation', *The Language of Psychoanalysis* (trans. D. Nicholson-Smith), London: Hogarth.

Lefebvre, H. (1991) *The Production of Space* (trans. D. Nicholson-Smith), Oxford: Blackwell.

Leopold, A. (1949) *A Sand County Almanac and Sketches Here and There*, Oxford University Press.

Leopold, A. (1987) '1947 Foreword', *Companion to* A Sand County Almanac: *Interpretive and Critical Essays* (ed. J. Baird Callicot), University of Wisconsin Press.

Leopold, A. (1992) 'Land pathology' (1935), *The River of the Mother of God and Other Essays* (ed. J. Baird Callicott and S. Flader), University of Wisconsin Press.

LeVay, S. (1993) *The Sexual Brain*, Cambridge, Massachusetts: The M.I.T. Press.

Lifton R. J. (1973) *Home From the War: Vietnam Veterans: Neither Victims nor Executioners*, New York: Simon and Schuster.

Lightfoot, P. (1981) *The Mekong*, Hove, East Sussex: Wayland.

Lloyd, G. E. R. (1970) *Early Greek Science: Thales to Aristotle*, London: Chatto & Windus.

Lockhart, G. (1989) *Nation in Arms: The Origins of the People's Army of Vietnam*, Sydney: Asian Studies Association of Australia/Allen and Unwin.

Long, N. V. (1973) *Before the Revolution: The Vietnamese Peasants Under the French*, Cambridge, MA: The M.I.T. Press.

Longinus (1965) 'On the sublime', *Classical Literary Criticism* (trans. T. S. Dorsch), Harmondsworth: Penguin.

Lucretius (1994) *On the Nature of the Universe* (trans. R. E. Latham, rev. J. Godwin), London: Penguin.

Lyotard, J. F. (1984) 'Answering the question: what is postmodernism?' (trans. R. Durand), in *The Postmodern Condition: A Report on Knowledge* (trans. G. Bennington and B. Massumi), Manchester University Press.

Lyotard, J. F. (1984) *Driftworks* (ed. R. McKeon), New York: Semiotext(e).

Lyotard, J. F. (1989) *The Lyotard Reader* (ed. A. Benjamin), Oxford: Basil Blackwell.

Lyotard, J. F. (1991) *The Inhuman: Reflections on Time* (trans. G. Bennington and R. Bowlby), Cambridge: Polity.

Lyotard, J. F. (1993) *Libidinal Economy* (trans. I. H. Grant), London: Athlone.

McAlister, J. T. Jr (1969) *Vietnam: The Origins of Revolution*, New York: Alfred A. Knopf.

MacCannell, J. F. (1991) *After the Patriarchy: The Regime of the Brother*, London: Routledge.

McComb, A. J. and P. S. Lake (1990) *Australian Wetlands*, North Ryde: Angus and Robertson.

McGarvey, B. E. (1988a) 'Landscape myths of the black swamp, Part One', *Northwest Ohio Quarterly*, 60, 2, pp. 57–68.

McGarvey, B. E. (1988b) 'Landscape myths of the black swamp, Part Two', *Northwest Ohio Quarterly*, 60, 3, pp. 95–104.

MacKenzie, J. M. (1988) *The Empire of Nature: Hunting, Conservation and British Imperialism*, Manchester University Press.

McKibben, W. (1990) *The End of Nature*, Harmondsworth: Penguin.

Maddock, K. (1978) 'Introduction', *The Rainbow Serpent: A Chromatic Piece* (ed. I. R. Buchler and K. Maddock), The Hague: Mouton.

Mahon, J. K. (1967) *History of the Second Seminole War 1835–1842*, University of Florida Press.

Marchant, L. (1982) *France Australe*, Perth: Artlook.

Markey, D. C. (1979) 'Pioneer Perth', *Western Landscapes* (ed. J. Gentilli), University of Western Australia Press.

Matthiessen, P. (1990) *Killing Mister Watson*, London: HarperCollins.

Meine C. (1988) *Aldo Leopold: His Life and Work*, University of Wisconsin Press.

Melville, H. (1990) *The Confidence-Man: His Masquerade* (ed. Stephen Matterson), Harmondsworth: Penguin.

Merchant, C. (1980) *The Death of Nature: Women, Ecology and the Scientific Revolution*, San Francisco: Harper & Row.

Michaels, W. B. (1977) '*Walden's* false bottoms', *Glyph*, 1, pp. 132–49.

Michigan State University Vietnam Advisory Group (no date), *My Thuan: A Mekong Delta Village in South Vietnam*, Washington, DC: Department of State Agency for International Development.

Miller, D. (1989) *Dark Eden: The Swamp in Nineteenth-Century American Culture*, Cambridge University Press.

Milne, A. A. (1958) *The World of Pooh Containing Winnie-the-Pooh and the House at Pooh Corner*, London: Methuen.

Mitchell, J. G. (1992) 'Our disappearing wetlands', *National Geographic*, 182, 4, pp. 3–45.

Mitsch, W. J. and J. G. Gosselink (1986) *Wetlands*, New York: Van Nostrand Reinhold.

Moore, G. F. (1884) *Diary of Ten Years Eventful Life of an Early Settler in Western Australia*, University of Western Australia Press.

Moore, T. (no date) 'A ballad. The Lake of the Dismal Swamp', *Poetical Works*, Edinburgh: Gall and Inglis.

Moretti, F. (1988) 'The spell of indecision', *Marxism and the Interpretation of Culture* (ed. C. Nelson and L. Grossberg), London: Macmillan.

Mountford, C. (1978) 'The Rainbow-Serpent myths of Australia', *The Rainbow Serpent: A Chromatic Piece* (ed. I. R. Buchler and K. Maddock), The Hague: Mouton.

Mudrooroo (1988) 'A snake story of the Nyoongah people: a children's tale', in *Longwater: Aboriginal Art and Literature Annual* (ed. U. Beier and C. Johnson), North Sydney: Aboriginal Artists Agency.

Mudrooroo (1994) 'Akurra serpent', *Aboriginal Mythology*, London: Harper-Collins.

Muir, J. (1987) *My First Summer in the Sierra* (1911), New York: Penguin.

Muir, J. (1991) *Our National Parks* (1901), San Francisco: Sierra Club Books.

Muir, J. (1992) *A Thousand-Mile Walk to the Gulf* (1916), New York: Penguin.

Mussolini, R. (1959) *My Life with Mussolini*, London: Robert Hale.

Nabokov, V. (1971) *Ada, or Ardor: A Family Chronicle* (1969), Harmonds-worth: Penguin.

Nash, R. (1982) *Wilderness and the American Mind* (3rd edn), Yale University Press.

Neumann, E. (1955) *The Great Mother: An Analysis of the Archetype* (trans. R. Manheim), London: Routledge and Kegan Paul.

Niering, W. A. (1966) *The Life of the Marsh: The North American Wetlands*, New York: McGraw Hill.

Niering, W. A. (1991) *The Wetlands of North America*, Charlottesville: Thomasson-Grant.

Noonuccal, O. and K. O. Noonuccal (1988) *The Rainbow Serpent*, Australian Government Publishing Service.

Oates, J. C. (1994) *Black Water*, London: Picador.

O'Connor, R., G. Quartermaine and C. Bodney (1989) *Report of an Investigation into Aboriginal Significance of Wetlands and Rivers in the Perth-Bunbury Region*, Western Australian Water Resources Council.

Oelschlaeger, M. (1991) *The Idea of Wilderness: From Prehistory to the Age of Ecology*, Yale University Press.

Paglia, C. (1991) *Sexual Personae: Art and Decadence from Nefertiti to Emily Dickinson*, New York: Vintage.

Paret, P. and J. W. Shy (1962) *Guerillas in the 1960s* (2nd edn), New York: Praeger.

Parker, R. (1983) *Miasma: Pollution and Purification in Early Greek Religion*, Oxford: Clarendon Press.

The Pentagon Papers (1971), New York: Quadrangle.

Pepys, S. (1972) *The Diary*, IV (1663) (ed. R. Latham and W. Matthews), London: G. Bell.

Pini, G. (1939) *The Official Life of Benito Mussolini* (trans. L. Villari), London: Hutchinson.

Plato (1954), 'Phaedo', in *The Last Days of Socrates* (trans. H. Tredennick), Harmondsworth: Penguin.

Pliny the Elder (1991) *Natural History: A Selection* (trans. J. F. Healy), London: Penguin.

Pollin, B. R. (1983) 'Edgar Allen Poe and John G. Chapman: their treatment of the Dismal Swamp and the Wissahickon', *Studies in the American Renaissance*, pp. 245–74.

Popham, D. (1980) *First Stage South: A History of the Armadale-Kelmscott District, Western Australia*, Town of Armadale.

'Porte Crayon' (1959) (D. H. Strother), 'The Dismal Swamp', *The Old South Illustrated* (ed. C. D. Eby), University of North Carolina Press.

Porter, R. (1994) 'Foreword', in A. Corbin, *The Foul and the Fragrant: Odour and the Social Imagination*, London: Picador.

Poynton, C. (1985) *Language and Gender: Making the Difference*, Deakin University Press.

Pugh, J. F. and F. T. Williams, *The Hotel in the Great Dismal Swamp and Contemporary Events Thereabouts*, Old Trap, NC: Jesse F. Pugh.

Purseglove, J. (1989) *Taming the Flood: A History and Natural of Rivers and Wetlands*, Oxford University Press.

Radcliffe-Brown, A. R. (1926) 'The Rainbow-Serpent myth of Australia', *Journal of the Royal Anthropological Institute of Great Britain and Ireland*, 56, pp. 19–25.

Radcliffe-Brown, A. R. (1930) 'The Rainbow-Serpent myth in South-East Australia', *Oceania*, 1, pp. 342–7.

Richardson, R. D. Jr (1986), *Henry Thoreau: A Life of the Mind*, University of California Press.

Richardson, R. D. Jr (1993) 'Introduction: Thoreau's broken task', in H. D. Thoreau, *Faith in a Seed*, Washington, DC: Island Press/Covelo, CA: Shearwater Books.

Robert, W. C. H. (1972) *The Explorations, 1696–1697, of Australia by Willem de Vlamingh*, Amsterdam: Philo.

Romein, J. (1978) 'Urbi et Orbi', *The Watershed of Two Eras: Europe in 1900* (trans. A. Pomerans), Wesleyan University Press.

Roosevelt, T. (1893) *The Wilderness Hunter*, New York: G. P. Putnam's Sons.

Ross, Sir D. (1995) *Aristotle* (6th edn), London: Routledge.

Rouch, H. (1987) 'La placenta comme tiers', *Langages*, 85, pp. 71–9.

Rubin, L. D. Jr (1989) *The Edge of the Swamp: A Study in the Literature and Society of the Old South*, Louisiana State University Press.

Ruskin, J. (1888) *Modern Painters*, II, Orpington, Kent: George Allen, 1888.

Ryan, M. and D. Kellner (1990) *Camera Politica: The Politics and Ideology of Contemporary Hollywood Film*, Indiana University Press.

Sanders, A. (1991) *Oral Histories Documenting Changes in Wheatbelt Wetlands*, Como, Western Australia: Department of Conservation and Land Management.

Sansom, R. L. (1970) *The Economics of Insurgency in the Mekong Delta of Vietnam*, Cambridge, MA: The M.I.T. Press.

Sartre, J.-P. (1969) *Being and Nothingness: An Essay on Phenomenological Ontology* (trans. H. E. Barnes), London: Methuen.

Schaffer, K. (1988) *Women and the Bush: Forces of Desire in the Australian Cultural Tradition*, Cambridge University Press.

Schama, S. (1988) *The Embarrassment of Riches: An Interpretation of Dutch Culture in the Golden Age*, University of California Press.

Schama, S. (1995) *Landscape and Memory*, London: HarperCollins.

Seddon, G. (1972) *Sense of Place: A Response to an Environment, The Swan Coastal Plain, Western Australia*, University of Western Australia Press.

Seddon, G. and D. Ravine (1986), *A City and its Setting: Images of Perth, Western Australia*, Fremantle Arts Centre Press.

Sesonske, A. (1982) 'Jean Renoir in Georgia: *Swamp Water*', *The Georgia Review*, 36, 1, pp. 24–66.

Seybert, A. (1799) 'Experiments and observations, on the atmosphere of marshes', *Transactions of the American Philosophical Society*, 4, pp. 415–30.

Sheldrake, R. (1991) *The Rebirth of Nature: The Greening of Science and God*, New York: Bantam.

Shelley, P. B. (1975) 'Alastor, or the Spirit of Solitude', *The Complete Poetical Works*, II (1814–1817) (ed. N. Rogers), Oxford: Clarendon Press.

Shepard, P. (1971) 'Ecology and man – a viewpoint', *The Everlasting Universe: Readings on the Ecological Revolution* (ed. L. J. Forstner and J. H. Todd), Lexington, MA: D. C. Heath.

Showalter, E. (1991) *Sexual Anarchy: Gender and Culture at the Fin de Siècle*, London: Penguin.

Simmel, G. (1950) 'The metropolis and mental life', *The Sociology of Georg Simmel* (trans. K. H. Wolff), New York: Macmillan.

Simms, W. G. (1853) 'The edge of the swamp', *Poems Descriptive, Dramatic, Legendary and Contemplative*, II, New York: Redfield.

Sjöö, M. and B. Mor (1991) *The Great Cosmic Mother: Rediscovering the Religion of the Earth* (2nd edn), San Francisco: Harper.

Sofoulis, Z. (1988) 'Through the Lumen: Frankenstein and the Optics of Re-Origination', Ph.D. thesis, History of Consciousness, University of California.

Sofoulis, Z. (1991) 'The return of the expressed: ethnocentrism in psychoanalytic cultural critique', Dismantle Fremantle Conference, June.

Sofoulis, Z. (1992) 'Hegemonic irrationalities and psychoanalytic cultural critique', *Cultural Studies*, 6, 3, pp. 376–94.

Soyinka, W. (1973) 'The swamp dwellers', *Collected Plays*, I, Oxford University Press.

Spengler, O. (1932) *The Decline of the West*, II (trans. C. F. Atkinson), London: George Allen and Unwin.

Stallybrass, P. and A. White (1986) *The Politics and Poetics of Transgression*, London: Methuen.

Stannage, C. T. (1979) *The People of Perth: A Social History of Western Australia's Capital City*, Perth City Council.

Stow, R. (1962) *To the Islands* (1958), Ringwood: Penguin.

Stowe, H. B. (1856) *Dred; A Tale of the Great Dismal Swamp*, Boston, MA: Phillips, Sampson.

Stratton-Porter, G. (1988) *A Girl of the Limberlost* (1910), New York: Signet.

Stratton-Porter, G. (1990) *Freckles* (1904), New York: Signet.

Styron, W. (1981) *The Confessions of Nat Turner* (1967), New York: Bantam.

Sussman, H. (1978) 'The deconstructor as politician: Melville's *Confidence-Man*', *Glyph*, 4, pp. 32–56.

Suzuki, D. (1991) 'Foreword', in W. J. Lines, *Taming the Great South Land: A History of the Conquest of Nature in Australia*, North Sydney: Allen and Unwin.

Swift, G. (1984) *Waterland*, London: Picador.

Tanham, G. K. (1967) *Communist Revolutionary Warfare: From the Vietminh to the Viet Cong* (rev. edn), New York: Praeger.

Theweleit, K. (1987) *Male Fantasies*, I: *Women, Floods, Bodies, History* (trans. S. Conway), Cambridge: Polity.

Theweleit, K. (1989) *Male Fantasies*, II: *Males Bodies: Psychoanalyzing the White Terror* (trans. C. Turner and E. Carter), Cambridge: Polity.

Thirsk, J. (1957) *English Peasant Farming. The Agrarian History of Lincolnshire from Tudor to Recent Times*, London: Routledge and Kegan Paul.

Thomas, K. (1984) *Man and the Natural World: Changing Attitudes in England, 1500–1800*, Harmondsworth: Penguin.

Thomas, W. (1976) *The Swamp*, New York: W. W. Norton.

Thompson, E. P. (1968) *The Making of the English Working Class* (1963), Harmondsworth: Penguin.

Thompson, L. (1966) 'Dismal Swamp', *Robert Frost: The Early Years, 1874–1915*, New York: Holt, Rinehart and Winston.

Thoreau, H. D. (1962) *Journal*, I–XIV (ed. B. Torrey and F. H. Allen) (1906), New York: Dover.

Thoreau, H. D. (1980) *Natural History Essays*, Salt Lake City: Gibbs Smith.

Thoreau, H. D. (1982) *The Portable Thoreau* (ed. C. Bode), New York: Penguin.

Thoreau, H. D. (1988) *The Maine Woods* (1864), New York: Penguin.

Thoreau, H. D. (1993) *Faith in a Seed: The Dispersion of Seeds and Other Late Natural History Writings* (ed. B. P. Dean), Washington, DC: Island Press/ Covelo, CA: Shearwater Books.

Threadgold, T. (1988) 'Language and Gender', *Australian Feminist Studies*, 6, pp. 41–70.

Tilden, F. (1970) *The National Parks*, New York: Alfred A. Knopf.

Tillyard, E. M. W. (1963) *The Elizabethan World Picture* (1943), Harmondsworth: Penguin.

Tragle, H. I. (1971) *The Southampton Slave Revolt of 1831: A Compilation of Source Material*, University of Massachusetts Press.

Tremayne, P. (1985) *Swamp*, New York: St Martin's Press.

Turnbull, D. (1989) *Maps are Territories, Science is an Atlas: A Portfolio of Exhibits*, HUS 203/204 Nature and Human Nature, Deakin University Press.

Turner, F. (1985) *Rediscovering America: John Muir in His Time and Ours*, New York: Viking.

Valenzuela, L. (1987) *The Lizard's Tail* (trans. G. Rabassa), London: Serpent's Tail.

Van Veen, J. (1949) *Dredge, Drain, Reclaim: The Art of a Nation*, The Hague: Trio.

Virilio, P. and S. Lotringer (1983) *Pure War* (trans. M. Polizotti), New York: Semiotext(e).

Waddy, B. B. (1975) 'Mosquitoes, malaria and man', in *Man-Made Lakes and Human Health* (ed. N. F. Stanley and M. P. Alpers), London: Academic Press.

Wallace, D. R. (1987) *Life in the Balance: Companion to the Audubon Television Specials*, San Diego: Harcourt Brace Jovanovich.

Washington, G. (1976) *Diaries*, I (ed. D. Jackson), University Press of Virginia.

Webb, G. G. (1847) 'Our Western Australian home; being sketches of scenery and society in the colony', *Swan River News and Western Australian Chronicle*, 44, pp. 160–2.

Williams J. and R. Abrashkin (1972) *Danny Dunn and the Swamp Monster*, London: Macdonald.

Williams, R. (1985) *The Country and the City* (1973), London: Hogarth.

Wittfogel, K. A. (1957) *Oriental Despotism: A Comparative Study of Total Power*, New Haven: Yale University Press.

Wohl, A. S. (1984) *Endangered Lives: Public Health in Victorian Britain*, London: Methuen.

Wolfe, L. M. (1978) *Son of the Wilderness: The Life of John Muir* (1945), University of Wisconsin Press.

Wolfe, L. M. (ed.) (1979) *John of the Mountains: The Unpublished Journal of John Muir* (1938), University of Wisconsin Press.

Wood, D. (1992) *The Power of Maps*, London: Routledge.

The World Commission on Environment and Development, *Our Common Future* (Australian edition), Melbourne: Oxford University Press.

Wright, J. (1990) 'Wilderness and wasteland', *Island*, 42, pp. 3–7.

Wright, T. (1799) 'On the mode most easily and effectually practicable of drying up the marshes of the maritime parts of North America', *Transactions of the American Philosophical Society*, 4, pp. 243–6.

Wynne, W. (1992) 'The Wetlands', *Lost Things and Other Poems*, Springwood, N.S.W.: Butterfly.

Index

WESTERN AUSTRALIAN WETLANDS
The Kimberley and South-West

Edited by Rod Giblett and Hugh Webb

Published by
Black Swan Press/
Wetlands Conservation Society

The first book to present the conservation values of the major wetlands in Western Australia is now available. The book is lavishly illustrated with 32 colour photographs of wetlands by Simon Neville, Jiri Lochman and others. It discusses all the wetlands in WA nominated to the Ramsar Convention on Wetlands of International Importance and includes maps of the wetlands discussed.

The book also includes a children's story about the Waugal by Mudrooroo, a history of wetlands' conservation in WA by Philip Jennings, a foreword by Christabel Chamarette and contributions by the editors on Aboriginal Country, the early history of Perth (including maps) and Mary Durack's classic *Kings in Grass Castles*. Poems, stories and statements from Aboriginal people about the significance of wetlands are interspersed throughout.

✂---

Please send me .... copies of *Western Australian Wetlands* @ A$20 each plus A$3 for postage and packaging. Enclosed please find a cheque for A$ ...... made out to Black Swan Press or please charge my Visa/Master/Bankcard (please circle):

Card number ....Expiry date ....

Signature ....Name on card ....

Name ....

Address ....

....

....

Please phone, fax, e-mail or post to: Black Swan Press, School of Communication and Cultural Studies, Curtin University of Technology, GPO Box U1987, Perth, Western Australia 6001. Tel. (09)3512253; fax (09)3517726; e-mail tbrownin@alpha2.curtin.edu.au

September 1996   192pp.   ISBN 1 86342 499 7